FRIED HERBERT MARTIN
FUNC MODELS
D1423174
40
QC174.45.A6.F89

FUNCTIONAL METHODS AND MODELS IN QUANTUM FIELD THEORY

H. M. Fried

The MIT Press
Cambridge, Massachusetts, and London, England

Printed and bound by Halliday Lithograph Corp.
in the United States of America

Library of Congress Cataloging in Publication Data

Fried, Herbert Martin.
Functional methods and models in quantum field theory.

1. Quantum field theory. I. Title.
QC174.45.F76 530.1'4 72-6969
ISBN 0-262-06047-7
ISBN 0-262-56011-9 (pbk.)

This volume is dedicated to the memory of

Yong Son Jin,

physicist, gentleman, colleague and friend.

PUBLISHER'S NOTE

The aim of this format is to close the time gap between the preparation of certain works and their publication in book form. A large number of significant though specialized manuscripts make the transition to formal publication either after a considerable delay or not at all. The time and expense of detailed text editing and composition in print may act to prevent publication or so to delay it that currency of content is affected.

The text of this book has been photographed directly from the author's typescript. It is edited to a satisfactory level of completeness and comprehensibility though not necessarily to the standard of consistency of minor editorial detail present in typeset books issued under our imprint.

The MIT Press

CONTENTS

Preface vii

List of Abbreviations ix

PART I: FUNCTIONAL METHODS 1

Chapter 1
INTRODUCTION 3

Chapter 2
THE GENERATING FUNCTIONAL AND THE S-MATRIX 14

A. The Generating Functional 14
B. Asymptotic Conditions 17
C. The S-Matrix 19
D. A Bremsstrahlung Example 23

Chapter 3
CONSTRUCTION OF THE GENERATING FUNCTIONAL 26

A. The Symanzik Construction 26
B. The Schwinger Construction 30
C. Several Interacting Fields 36
D. Rearrangements/Grouping of Feynman Graphs 42
E. Fields at the Same Point 49

Chapter 4
NONCANONICAL (CHIRAL) GENERALIZATIONS 59

Chapter 5
SPECIAL TOPICS IN QUANTUM ELECTRODYNAMICS 68

A. The Heavy Proton Limit 68
B. Green's Function Equations 73
C. Gauge Transformations and the Ward Identity 77

PART II: MODEL APPROXIMATIONS 89

Chapter 6
PERTURBATION EXPANSIONS 91

A. Connectedness and Irreducibility 91
B. The Born (Tree Graph) Functional 93
C. Lowest Order Radiative Corrections 97
D. Renormalization Procedures 114

Chapter 7
SOLUBLE MODELS 119

A. Two-dimensional Electrodynamics 119
B. The Thirring Model 125
C. Neopolitan Models 128
D. The Lee Model 130

Chapter 8
NO-RECOIL METHODS 132

A. The Bloch-Nordsieck Approximation 132
B. Soft Photons: Fermion Self-Energy Structure 139
C. Soft Photons: Cancellation of Infrared Divergences 142
D. Soft Pions 149

Chapter 9
RELATIVISTIC EIKONAL PHYSICS 153

A. Small-Angle Formalism 154
B. Wide-Angle Approximations 163
C. Inelastic (Bremsstrahlung) Models 169

Chapter 10
SPECULATIONS AT HIGH ENERGY 175

A. Multiperipheral Field Theory 175
B. Summing Leading Logs 182
C. A Limiting Model 194

THE APPENDIX 202

The purpose of this slim volume is to present from the same, unified point of view a description of the major soluble and approximate models of relativistic quantum field theory. While the content of realistic field theories remains a matter of speculation, it seems to be quite clear that an understanding, deeper than phenomenological, of the structure of matter waits upon an understanding of the dynamics of interacting fields. For this reason alone it is worthwhile to set down those results which are known about model theories. Considerations of efficiency, together with the anticipation of future usefulness, suggest the employment of the beautiful functional methods developed by Schwinger, Feynman, Symanzik and many others.

It should perhaps be stressed that this is not intended to be a small treatise on the quantum theory of fields. There are many readable, several excellent, such works now available, and there is little point in adding another to the list; duplication of readily available material is not desired. The object of these remarks is to discuss those models of quantum field theory known at present, and to anticipate those to come; and with a few omissions this goal can be realized most efficiently with the aid of functional methods.

The level of presentation attempted here is such that students familiar with conventional field theoretic arguments should find the transition to a functional description quite painless, and perhaps even a little exciting. Those already familiar with the methods may find these remarks helpful as a repository of well-known models, together with a collection of useful tricks of the functional trade. The latter comprise the major portion of the Appendix, while the main part of the text is divided into two sections, the first dealing with functional preliminaries, and the second with the models themselves.

Most of this material was culled from lectures given by the author at Brown University during the academic years 1965-70. It is both necessary and proper that acknowledgment should here be made to those persons from whom, over the previous half-decade, the author had the opportunity and pleasure of learning both physics and techniques: Professors J. Schwinger, K. Symanzik, and B. Zumino. Essentially all of the material of Part I has appeared in their published papers or unpublished lecture notes during this period. In addition, it is most fitting to acknowledge helpful and informative conversations with colleagues at Brown and elsewhere; thanks are particularly due to Prof. R. Blankenbecler for many discussions on topics covered in the latter chapters; and to Prof. K. Johnson for a reading of the manuscript. In no small measure is the author indebted to Prof. S. Drell for extending the hospitality of SLAC over several summers, where many of the functional applications of Part II were studied; and to the A. E. C. for its generous contractual support of high energy physics at Brown. Special thanks are due to Judith Bardsley, of the Engineering Division at Brown, for her most efficient and elegant typing of the manuscript.

It may be appropriate to paraphrase the definition of a Perfect Seminar given by Prof. F. Bloch during one of his superb lectures at Stanford some two decades ago: One which begins in so simple a manner that everyone in the audience understands all of the speaker's remarks, but which gradually becomes more difficult and less comprehensible to most of the listeners, until at the very end of the lecture not even the speaker understands what he is talking about. Much the same spirit governs the presentation of the latter chapters of this work, especially the last. Nevertheless, and in spite of the avoidance of any mathematical rigor, it is hoped that this material will aid in again making field theory more relevant to experimental physics.

H. M. Fried
Brown University

LIST OF ABBREVIATIONS

The following symbols have been used freely throughout the text.

ETCR
Equal-time commutation relations

QED
Quantum electrodynamics

CM
Center-of-mass

NVM
Neutral vector meson

UV
Ultraviolet

IR
Infrared

LHS
Left-hand side

RHS
Right-hand side

PART I

FUNCTIONAL METHODS

Chapter 1

INTRODUCTION

Quantum field theory may be thought of as a theory of coupled Green's functions which, according to some specific interaction, describe all possible scattering and production processes undergone by particles, the quanta of the fields under consideration. Operator field equations, together with the kinematical restrictions of equal time commutation relations, may be used to represent the infinite set of coupled equations for these functions. From knowledge of these Green's functions one may pass, by the process of renormalization, to a description of the interacting particles of the particular theory under consideration.

The objective of any field theory is a description of particles, given in terms of a set of underlying, basic fields. The reason for this indirect approach is that physical quantities, such as energy and angular momentum, conserved during the entire course of any interaction involving annihilation, scattering and production of particles, are far more readily described in terms of fields than in terms of the particles themselves. This situation arises from the admixture of quantum mechanics with relativity: any description of a system over a finite time interval Δt necessarily involves an uncertainty in the energy of that system, of amount $\Delta E \gtrsim h/\Delta t$; and since ΔE may be interpreted as $c^2\Delta m = c^2 m\Delta n$, where m denotes the mass of an appropriate particle and Δn measures the number of such particles, the shorter the time interval the larger the number of virtual particles which must be included. However, the field variables can be grouped into combinations whose conservation may be simply exhibited at all times, while the identification of these conserved quantities in terms of particle properties is performed asymptotically. These ideas have been emphasized with great persuasiveness by Schwinger, over the past two decades, and they form the conceptual foundation for essentially all the material to follow.

In practice, mathematical complexities have prevented the realization of this program in realistic situations. Thus, the restriction to perturbation approximations contains the tacit assignment of one field to each particle. Non-perturbative methods are clearly indicated as a proper, if not imperative, subject for intensive study, and one step in that direction is the examination of the known approximate, or model, field theories. Perhaps the most efficient way of displaying both the complexities of the problem and the small amount of understanding gleaned from the soluble models, is with the aid of functional methods. These methods, at least in the context employed here, may be considered as providing a realization of the coupled Green's function equations, in the sense that the latter may be reconstructed from the formal functional solutions exhibited for any given interaction. In addition, there are many non-perturbative approximations, partial sums of limited Feynman graphs corresponding to models of various sorts, which are suggested by the formalism, and which one can attempt to apply to various physical situations.

The initial step is the specification of a Lagrangian. One can then follow Schwinger[1] and adopt a single quantum-mechanical action principle to obtain, in essence, differential equations for desired amplitudes; or one can follow Feynman[2] and write the integral form of Schwinger's action principle in terms of functional integrals; or one may follow Symanzik's procedure[3] of constructing a generating functional from the separately stated field equations and equal time commutation relations. Essentially this latter route will be taken here, although the final expressions will resemble Schwinger's and can easily be cast into Feynman's form. A very brief discussion of the action principle formalism, which does not nearly do justice to the clarity and beauty of the original presentation, will be included in order to permit a comparison of the results of these different approaches.

It will be useful to group together certain well-known properties of free boson and fermion fields, for use in subsequent chapters. Individual field equations which result from the variation of specific Lagrangians will be written when needed. A more complete discussion of Lagrangians and their variations may be found in numerous texts;[4] only those few points directly relevant to the functional procedures will be employed here.

The simplest example is that of the scalar boson free-field, which satisfies the Klein-Gordon field equation[5]

$$(m^2-\partial^2)A(x) = K_x A(x) = 0, \tag{1.1}$$

and the equal time commutation relations

$$[A(x),A(y)]\Big|_{x_o=y_o} = 0, \tag{1.2}$$

$$[A(x),\dot{A}(y)]\Big|_{x_o=y_o} = i\delta(\overrightarrow{x-y}). \tag{1.3}$$

The same equal time commutation relations will also be assumed for the interacting (unrenormalized) fields.[6] The reason for this is that such relations are basically kinematical statements, giving the fundamental second-quantization rule between generalized coordinates and momenta. The motion of that particle of mass m, associated with the free field A, is described by the causal propagator

$$\Delta_c(x-y) = i\langle\big(A(x)A(y)\big)_+\rangle \tag{1.4a}$$

$$= (2\pi)^{-4}\int d^4k\, e^{ik\cdot x}(k^2+\mu^2-i\varepsilon)^{-1}\Big|_{\varepsilon\to 0+} \tag{1.4b}$$

$$= \frac{i\mu}{4\pi^2}\frac{\theta(x^2)}{\sqrt{x^2}}K_1(\mu\sqrt{x^2}) - \frac{\mu}{8\pi}\frac{\theta(-x^2)}{\sqrt{-x^2}}H_1^{(2)}(\mu\sqrt{-x^2})$$

$$+ \frac{1}{4\pi}\delta(x^2), \tag{1.4c}$$

where $\langle\cdots\rangle$ denotes vacuum expectation value, and $(\ )_+$ refers to a time ordering of the operators, the latest standing to the left. The simplest method of obtaining this dependence, which will be useful in subsequent manipulations, is to take advantage of the possibility of decomposing a free-field operator into a sum of positive and negative frequency parts,

$$A(x) = A^{(+)}(x) + A^{(-)}(x),$$

$$A^{(+)}(x) = (2\pi)^{-3/2} \int d^3k(2\omega)^{-1/2} a(\vec{k}) e^{i\vec{k}\cdot\vec{r}-i\omega x_o},$$

$$A^{(-)}(x) = (2\pi)^{-3/2} \int d^3k(2\omega)^{-1/2} a^\dagger(\vec{k}) e^{-i\vec{k}\cdot\vec{r}+i\omega x_o},$$

where $\omega = [\vec{k}^2+\mu^2]^{1/2}$, and the destruction $(a(\vec{k}))$ and creation $(a^\dagger(\vec{k}))$ operators satisfy the commutation relations

$$[a(k),a(k')] = 0, \quad [a(k),a^\dagger(k')] = \delta(\vec{k}-\vec{k}').$$

Hence, the commutators

$$[A^{(+)}(x),A^{(-)}(y)] \equiv i\Delta^{(+)}(x-y) \equiv -i\Delta^{(-)}(y-x) \qquad (1.5)$$

are c-numbers, with the value

$$\Delta^{(\pm)}(x) = \mp i(2\pi)^{-3} \int d^4k\delta(k^2+\mu^2)\theta(\pm k_o)e^{ik\cdot x} \qquad (1.6)$$

where $k\cdot x \equiv \vec{k}\cdot\vec{r}-k_o x_o$ and $\theta(z) = 1$ for $z > 0$, and $= 0$ for $z < 0$.

Because their Fourier transforms are on the mass shell, that is, are proportional to $\delta(k^2+\mu^2)$, both $\Delta^{(\pm)}$ satisfy the homogeneous Klein-Gordon equation (1.1), as do their frequently convenient linear combinations

$$\Delta(x) = \Delta^{(+)}(x) + \Delta^{(-)}(x)$$

and

$$-i\Delta_{(1)}(x) = \Delta^{(+)}(x) - \Delta^{(-)}(x).$$

It should be noted that $\Delta(x)$ vanishes for spacelike values of its argument, which is the statement of microscopic causality for the operator fields.

Solutions of the inhomogeneous Klein-Gordon equation,

$$(\mu^2-\partial^2)\Delta_c(x) = \delta^4(x), \qquad (1.7)$$

may be constructed out of $\Delta^{(\pm)}$ using the definition (1.4), equivalent to the relation

$$\Delta_c(x) = -\theta(x_o)\Delta^{(+)}(x) + \theta(-x_o)\Delta^{(-)}(x). \qquad (1.8a)$$

There are four such solutions to (1.7), denoted by Δ_c, $\Delta_{\bar{c}}$, Δ_R, Δ_A, and described as the causal, anticausal, retarded and advanced Green's functions, respectively. These functions differ fundamentally in their asymptotic (in time) properties, as is easily seen in terms of the definitions (1.8a) and

$$\Delta_{\bar{c}}(x) = \Delta_c^*(x) = -\theta(x_o)\Delta^{(-)}(x) + \theta(-x_o)\Delta^{(+)}(x_o), \qquad (1.8b)$$

$$\Delta_R(x) = -\theta(x_o)\Delta(x), \qquad (1.8c)$$

$$\Delta_A(x) = +\theta(-x_o)\Delta(x), \qquad (1.8d)$$

while the limitation to four such solutions is perhaps seen most easily from the corresponding Fourier transforms of these functions,

$$\tilde{\Delta}_c(k) = (k^2+\mu^2-i\varepsilon)^{-1}, \quad \text{as in (1.4a)},$$

$$\tilde{\Delta}_{\bar{c}}(k) = (k^2+\mu^2+i\varepsilon)^{-1},$$

$$\tilde{\Delta}_R(k) = \left(k^2+\mu^2-i\varepsilon\cdot\varepsilon(k_o)\right)^{-1},$$

$$\tilde{\Delta}_A(k) = \left(k^2+\mu^2+i\varepsilon\cdot\varepsilon(k_o)\right)^{-1},$$

with $\varepsilon(k_o) \equiv \theta(k_o) - \theta(-k_o)$.

In a similar fashion, the field equation and commutation relations for the electromagnetic four-vector potential (in the Feynman gauge) may be taken to be those given above if an extra factor of $\delta_{\mu\nu}$ is appended to the right-hand side of each expression, in order to match the replacement $A \to A_\mu$. This carries with it the well known restrictions of indefinite metric; but all of our subsequent electrodynamic Green's functions will be generalized to arbitrary relativistic gauges, could have been written down in the fundamental radiation gauge, and hence this substitution is just a simplifying artifice. The point is that one may avoid introducing the indefinite metric, which corresponds to a change in the ground rules of quantum mechanics, by remaining within the radiation gauge, where one has a gauge invariant process of second quantization, with multipoint Green's functions whose properties follow from radiation gauge operators. Lorentz covariance is not explicit, but may be demonstrated by adjoining operator gauge transformations to every Lorentz transformation. For the calculation of physical effects, it is far simpler to first make a transformation out of the radiation gauge and into one of the relativistic gauges, and to perform one's computations there. In particular, one could go to the Feynman gauge, which is just where one would arrive by taking the simpler route of quantization by indefinite metric.

Useful statements for free Fermion fields are the Dirac equations,

$$\mathcal{D}_x\psi(x) = (m+\gamma\cdot\partial_x)\psi(x) = 0, \tag{1.9a}$$

$$\bar{\psi}(x)\overleftarrow{\bar{\mathcal{D}}}_x = \bar{\psi}(x)(m-\gamma\cdot\overleftarrow{\partial}_x) = 0, \tag{1.9b}$$

and the equal time anticommutation relations

$$\{\psi_\alpha(x),\bar{\psi}_\beta(y)\}\Big|_{x_o=y_o} = \gamma_4^{\alpha\beta}\delta(\overrightarrow{x-y}), \tag{1.10a}$$

$$\{\psi_\alpha(x),\psi_\beta(y)\}\Big|_{x_o=y_o} = \{\bar{\psi}_\alpha(x),\bar{\psi}_\beta(y)\}\Big|_{x_o=y_o} = 0. \tag{1.10b}$$

Equations (1.10) will also be assumed for the interacting fermion fields.[6] The causal propagator for a free fermion of mass m may be written as

$$S_c(x-y) = i\langle(\psi(x)\bar{\psi}(y))_+\rangle \tag{1.11a}$$

$$= (m-\gamma\cdot\partial_x)\Delta_c(x-y;m^2), \tag{1.11b}$$

where the symbol $(\ )_+$ now includes an extra factor $(-1)^P$, with P representing the number of permutations of the fermion fields from the ordering displayed. Explicit forms for (1.11) follow from (1.4).

Just as in the boson case, one may split the free fermion fields into positive and negative frequency parts, $\psi = \psi^{(+)} + \psi^{(-)}$, $\bar{\psi} = \bar{\psi}^{(+)} + \bar{\psi}^{(-)}$, with the representations

$$\psi(x) = (2\pi)^{-3/2}\sum_{s=1}^{2}\int d^3p\left[\frac{m}{E}\right]^{1/2}\{b_s(\vec{p})e^{ip\cdot x}\,u_s(p)$$

$$+ d_s^\dagger(\vec{p})e^{-ip\cdot x}\,v_s(p)\}$$

$$\bar{\psi}(x) = (2\pi)^{-3/2}\sum_{s=1}^{2}\int d^3p\left[\frac{m}{E}\right]^{1/2}\{b_s^\dagger(\vec{p})e^{-ip\cdot x}\,\bar{u}_s(p)$$

$$+ d_s(\vec{p})e^{ip\cdot x}\,\bar{v}_s(p)\},$$

where $E = [\vec{p}^2+m^2]^{1/2}$, $p\cdot x = \vec{p}\cdot\vec{r}-Ex_o$ and $\bar{u} = u^\dagger\gamma_4$, $\bar{v} = v^\dagger\gamma_4$. The $b_s(\vec{p})$ and $b_s^\dagger(\vec{p})$ are destruction and creation operators for fermions of spin index s and momentum $\vec{p}$, while $d_s(\vec{p})$ and $d_s^\dagger(\vec{p})$ play the corresponding role for the antifermions. The spinors $u_s^\alpha(p)$ and $v_s^\alpha(p)$ satisfy the normalization conditions

$$\sum_\alpha \bar{u}_s^\alpha(p) u_{s'}^\alpha(p) = \delta_{ss'},$$

$$\sum_\alpha \bar{v}_s^\alpha(p) v_{s'}^\alpha(p) = -\delta_{ss'}, \tag{1.12}$$

$$\sum_\alpha \bar{u}_s^\alpha(p) v_{s'}^\alpha(p) = 0,$$

with the sum running over the four values of the Dirac component index α; and they also possess the closure properties

$$\delta_{\alpha\alpha'} = \sum_{s=1}^{2} \{u_s^\alpha(p)\bar{u}_s^{\alpha'}(p) - v_s^\alpha(p)\bar{v}_s^{\alpha'}(p)\}. \tag{1.13}$$

Both solutions $S^{(\pm)}$ to the homogeneous Dirac equation, as well as all four solutions S_c, $S_{\bar{c}}$, S_R, S_A to the inhomogeneous Dirac equation can be represented most succinctly by the relation (1.11b), with each fermion function given by the application of $(m-\gamma\cdot\partial)$ to the corresponding boson function.

From the input information of field equations plus equal time commutation or anticommutation relations, one may construct formal solutions for the variety of time-ordered n-point Green's functions,

$$<\left(\psi(x_1)\cdots\psi(x_\ell)\bar{\psi}(y_1)\cdots\bar{\psi}(y_m)A(z_1)\cdots A(z_p)\right)_+>. \tag{1.14}$$

The boson fields appearing in (1.14) are assumed to commute with all fermion fields at equal times; two or more kinematically independent fermion fields will be assumed to anticommute at equal times. Each of these n-point functions is, by virtue of translational invariance, a function of $n-1$ coordinate differences. For $n = 2$ one builds the propagators, the dressed generalizations of (1.4) and (1.11), whose behavior for large values of coordinate separation yields information about the excitations, or quanta, of the coupled fields. Scattering and production of these particles are described by the amputated, mass shell, Fourier transforms of those Green's functions with $n = 4$ and $n \geq 5$, respectively, while the vertex function, $n = 3$, plays an important role in the evaluation of these quantities.

Elaboration and illustration of these remarks will be made within the context of the various soluble models and approximations, but it may be useful to comment on the most striking aspect of any such theory of interacting fields: one finds, in general, a change in the mass and coupling parameters, m_0 and g_0, originally introduced in the Lagrangian. If a given field is to have associated with it a particle of mass m, the dressed propagator of that field will have, in momentum space, a pole at m^2, where $m \neq m_0$, together with a nonzero residue, Z, at that pole. Such Z-factors, the wave function renormalization constants, combine multiplicatively with g_0, and with related constants, to provide a renormalized coupling constant, g; and all physical effects are then to be expressed in terms of m and g. Because of our present inability to solve, or to approximate in a satisfactory way, the coupled Green's function equations, one assumes that solutions exist which do correspond to finite, physical masses and measured coupling constants, and one then proceeds to calculate fluctuation effects in terms of such physical parameters.

Renormalizable theories are characterized by perturbative expansions whose degree of divergence does not increase with the order of the expansion, and conversely for nonrenormalizable theories; in the former case, all divergences can be absorbed in the transition from m_0 and g_0 to m and g. It is not known, at present, whether the bare parameters are really infinite, as suggested by the perturbation approximations, or if such divergences are due to that method of calculation.

In order to treat strongly interacting particles, nonperturbative approximations to the coupled field equations are clearly necessary; and here one is at a conceptual disadvantage compared to the perturbative case where one certainly expects the simplest or lowest order approximation to yield the dominant contribution to any physical effect so calculated. One must always specify a parameter which is to be treated as "small" when defining an approximation, and to do this one should have an at least intuitive understanding of which quantities are "large" and which are not. Fortunately, there is a large and growing body of literature and experiments which suggest that the eikonal models of Part II act as a first approximation, providing a qualitative field theoretic description of some very high energy particle processes.

Notes

1. J. Schwinger, Proc. Nat. Acad. of Sciences (USA), 37, 452 (1951); and Harvard Lectures (1954).

2. R. P. Feynman and A. R. Hibbs, *Quantum Mechanics and Path Integrals*, McGraw-Hill Book Company (New York: 1965).

3. K. Symanzik, Z. Naturforschung, 9a, 10 (1954), 809.

4. S. Schweber, *An Introduction to Relativistic Quantum Field Theory*, Row-Peterson & Co. (New York: 1961); N. N. Boguliubov and D. V. Shirkov, *Introduction to the Theory of Quantized Fields*, Interscience Publishers, Inc. (New York: 1959); J. Bjorken and S. Drell, *Relativistic Quantum Fields*, McGraw-Hill Book Company (New York: 1965).

5. The relativistic notation used throughout is such that

$a_\mu = (\vec{a}, ia_o)$, $a \cdot b = \vec{a} \cdot \vec{b} - a_o b_o$, $\partial^2 = \vec{\nabla}^2 - \partial_o^2$,

$\gamma_\mu^\dagger = \gamma_\mu, \quad \{\gamma_\mu, \gamma_\nu\} = 2\delta\mu\nu.$

6. It is not at all clear that these ETCR apply to the interacting situation. Equations for the propagator of every fully-interacting field, constructed in part from the assumptions (1.3) and (1.10), must have solutions which display asymptotic properties consistent with these assumptions. For example, one expects the behavior of $[A(\vec{x},t),A(\vec{y},0)]$ as $t \to 0$ to be characterized by eigenstates of the total Hamiltonian with asymptotically large eigenvalues. For nonrelativistic systems, these states are the same as those of the free-field Hamiltonian, and hence there is no inconsistency in the use of the same ETCR. In an interacting, relativistic field theory, the possibility always remains that the behavior of such quantities as t vanishes is different from that originally assumed and expected, and one can have little confidence in assumptions concerning Green's functions of asymptotic momenta. The author is indebted to Professor K. Johnson for a discussion of this point.

Chapter 2

THE GENERATING FUNCTIONAL AND THE S-MATRIX

The probability amplitudes, or S-matrix elements, which describe transitions between states of physical systems can be presented in the form of certain mass shell operations upon appropriate multipoint Green's functions, (1.14). The latter can be most succinctly obtained from a generating functional, which will, in turn, permit the so-called reduction formulae of the S-matrix to be exhibited in a clear and compact way.

A. The Generating Functional

It is simplest to begin by considering[1] a scalar, hermitian, boson field $A(x)$; the generalization to other fields may easily be performed at a later stage. A "generating functional operator," labeled by two space-like surfaces σ_a and σ_b, is defined by

$$\overset{a}{T^{b}}\{j\} = \left(e^{i\int_b^a jA}\right)_+, \tag{2.1}$$

where $\int_b^a jA \equiv \int_{\sigma_b}^{\sigma_a} d^4x j(x)A(x)$, with $j(x)$ denoting an arbitrary, c-number source function. The surfaces $\sigma_{a,b}$ may be thought of as flat time-cuts, in which case the integral of (2.1) becomes $\int d^3x \int_{t_b}^{t_a} dt j(x)A(x)$, where the volume integral is understood to extend over all space, and where $t_a \geq t_b$. The time ordering prescription of (2.1) means that every term in the expansion of the exponential is to be arranged so that the later operators stand to the left. (Ordered exponentials are discussed in some detail in Appendix A.)

The functional derivative (Appendix B) of $\overset{a}{T^{b}}$ is given by

$$\frac{\delta}{\delta j(x)} \overset{a}{\underset{b}{T}}\{j\} = i\left(A(x)e^{i\int_b^a jA}\right)_+, \tag{2.2}$$

$$= i\overset{a}{\underset{x}{T}}A(x)\overset{x}{\underset{b}{T}} \tag{2.3}$$

if the point x lies on a space-like surface between σ_a and σ_b (or if $t_a > x_o > t_b$), and is zero otherwise. Equation (2.3) is a sometimes useful way of expressing the ordering requirements of (2.2).

Variations of $\overset{a}{\underset{b}{T}}$ due to an infinitesimal change of the surface σ_a, about the point x_a, may be written in the form

$$\delta\overset{a}{\underset{b}{T}} = i\delta\sigma(x_a)\cdot j(x_a)\left(A(x_a)e^{i\int_b^a jA}\right)_+,$$

or

$$\frac{\delta\overset{a}{\underset{b}{T}}}{\delta\sigma(x_a)} = ij(x_a)A(x_a)\overset{a}{\underset{b}{T}}, \tag{2.4}$$

where the second line of (2.4) follows from (2.3) and the property $\overset{c}{\underset{c}{T}} = 1$. Repeating these steps for a small variation of σ_a about the point y_a yields

$$\frac{\delta}{\delta\sigma(y_a)}\frac{\delta}{\delta\sigma(x_a)}\overset{a}{\underset{b}{T}} = i^2 j(x_a)j(y_a)A(x_a)A(y_a)\overset{a}{\underset{b}{T}}. \tag{2.5}$$

Had these variations been performed first at y_a and then at x_a, the result would have been (2.5) with the labels interchanges; hence, the difference of these procedures gives

$$\left[\frac{\delta}{\delta\sigma(x_a)}, \frac{\delta}{\delta\sigma(y_a)}\right]\overset{a}{\underset{b}{T}} = i^2 j(x_a)j(y_a)[A(x_a),A(y_a)]\overset{a}{\underset{b}{T}}. \tag{2.6}$$

If the points x_a, y_a are on the same space-like surface, as assumed, the left side of (2.6) must vanish, and this means that the commutator $[A(x),A(y)]$ must vanish spacelike. Thus, microcausality is a necessary condition for the integrability of $\overset{a}{T}_b$.

The adjoint functional operator $\overset{a\dagger}{T}_b$ may be written as

$$\overset{a\dagger}{T}_b\{j\} = \left(e^{-i\int_b^a jA}\right)_-, \tag{2.7}$$

where $(\;)_-$ denotes a reversed time ordered bracket. With (2.7), it is simple to demonstrate that $\overset{a}{T}_b$ is a unitary operator, which property is essential if the S-matrix is to be expressed in terms of $\overset{+\infty}{T}_{-\infty}$. Because one has, immediately,

$$\frac{\delta}{\delta\sigma(x_a)}\left(\overset{a\dagger}{T}_b\overset{a}{T}_b\right) = 0, \tag{2.8}$$

the combination $\overset{a\dagger}{T}_b\overset{a}{T}_b$ is independent of σ_a; hence it can be evaluated at any σ_a, in particular $\sigma_a = \sigma_b$, which provides the relation $\overset{a\dagger}{T}_b\overset{a}{T}_b = 1$. By considering variations of a point on the surface σ_b, one may prove in the same manner that $\overset{a}{T}_b \cdot \overset{a\dagger}{T}_b = 1$, which completes the demonstration of unitarity. The greater part of the considerations to follow shall be concerned with that operator obtained by letting the space-like surfaces σ_a and σ_b recede to $+\infty$ and $-\infty$, respectively, and this quantity, $\overset{+\infty}{T}_{-\infty}$, will be written simply as T.

The usefulness of $T\{j\}$ is that it permits one to construct all products of time-ordered operators, by performing a suitable number of functional differentiations with respect to the source $j(x)$, and then setting the source equal to zero. Only the vacuum expectation values of such products are actually needed, and for this one defines the generating functional $Z\{j\} = \langle T\{j\}\rangle$. More generally, one should include other fields in T and Z in order to be able to construct n-point Green's functions of the form of (1.14). Fermions may be introduced with the aid of anticommuting c-number sources, $\eta_\alpha(x)$, $\bar{\eta}_\beta(y)$, which anticommute with themselves and all fermion fields ψ, $\bar{\psi}$. In contrast to the natural property valid for the boson sources,

$$\left[\frac{\delta}{\delta j(x)}, j(y)\right] = \delta(x-y), \tag{2.9}$$

the fermion sources are to satisfy[2]

$$\left\{\frac{\delta}{\delta \eta_\alpha(x)}, \eta_\beta(y)\right\} = \left\{\frac{\delta}{\delta \bar{\eta}_\alpha(x)}, \bar{\eta}_\beta(y)\right\} = \delta_{\alpha\beta}\delta(x-y), \tag{2.10}$$

with all other combinations anticommuting,

$$\left\{\frac{\delta}{\delta\eta}, \frac{\delta}{\delta\bar{\eta}}\right\} = \left\{\frac{\delta}{\delta\eta}, \frac{\delta}{\delta\eta}\right\} = \left\{\frac{\delta}{\delta\eta}, \psi\right\} = \left\{\frac{\delta}{\delta\eta}, \bar{\psi}\right\} = 0. \tag{2.11}$$

Then, functional differentiation of the generating functional operator

$$T\{j,\eta,\bar{\eta}\} = \left(e^{i\int[jA+\bar{\psi}\eta+\bar{\eta}\psi]}\right)_+ \tag{2.12}$$

will produce the products (1.14) with the correct change of sign under fermion permutation, as in (1.11). The complete generating functional for scalar bosons and fermions is then taken as $Z\{j,\eta,\bar{\eta}\} = \langle T\{j,\eta,\bar{\eta}\}\rangle$.

B. Asymptotic Conditions

As yet nothing has been said about the nature of the interaction coupling the fields A and $\psi,\bar{\psi}$, since it is only necessary to assume that particle-like solutions exist to the coupled Green's function equations in order to define the S-matrix. It shall be here assumed that the Fourier transforms of both fermion and boson propagators have, as a function of their appropriate invariant variable, a single pole occurring at the appropriate physical masses, with residues Z_2 and Z_3, respectively. An equivalent physical statement is that, in the remote past and future, one is dealing with particles sufficiently separated to be essentially noninteracting; and this suggests that the fields of interest can, in such asymptotic regions, be replaced by free fields describing free particles carrying their observed, physical masses. These are the IN- and OUT- fields of Yang and Feldman,[3] which are conventionally related to the asymptotic, unrenormalized field operators by the postulates of the "weak asymptotic condition,"

$$\lim_{x_0 \to \mp\infty} [\langle a|A(x)|b\rangle - \sqrt{Z_3}\, \langle a|A_{\substack{IN\\OUT}}(x)|b\rangle] = 0, \tag{2.13a}$$

$$\lim_{x_0 \to \mp\infty} [\langle a|\psi(x)|b\rangle - \sqrt{Z_2}\, \langle a|\psi_{\substack{IN\\OUT}}(x)|b\rangle] = 0, \tag{2.13b}$$

where $|a\rangle$, $|b\rangle$ are arbitrary, time-independent states of the system under consideration, specified by the sets of quantum numbers a,b. Unless otherwise noted, we shall always use a complete set of IN-particle states, $|a\rangle = |a\rangle_{IN}$; in particular, the vacuum state used to define $Z\{j,\eta,\bar{\eta}\}$ is constructed with the IN-vacuum state, $\rangle = |0\rangle_{IN}$. These states are constructed in terms of Fourier transforms of products of $A_{\substack{IN\\OUT}}$ operators acting upon the vacuum states $|0\rangle_{\substack{IN\\OUT}}$, with the latter defined by the limiting operations of time-dependent (Schroedinger picture) states,

$$|0\rangle_{\substack{IN\\OUT}} = \lim_{t \to \mp\infty} |0,t\rangle, \tag{2.14}$$

with all external c-number sources set equal to zero. Equation (2.13) must be written as a relation between matrix elements in order to prevent the contradiction which would appear if a strong asymptotic condition were assumed; by the latter statement one means an operator relation such as $\lim_{t \to -\infty} [A(x) - \sqrt{Z_3}\, A_{IN}(x)] = 0$, which has, as an immediate consequence (Appendix C), the incorrect relation $Z_3 = 1$, as for free fields. The presence of the $Z_{2,3}$ in (2.13) is a reflection of the self-interaction which the particles must undergo, even when they are sufficiently separated so that there is little interaction between them; and it is the removal of these factors, along with a mass shift generated by the same self-interaction, which corresponds to the process of renormalization. Another and somewhat more convenient way of writing (2.13) is in the form

$$A(x) = \sqrt{Z_3}\, A_{\substack{IN \\ OUT}}(x) + \int d^4y \Delta_{\substack{R \\ A}}(x-y) K_y A(y), \tag{2.15}$$

with (2.15) always to be understood as a relation between matrix elements.

C. The S-Matrix

Conventionally defined as that unitary operator which transforms IN states and operators into OUT states and operators,

$$A_{OUT}(x) = S^{\dagger} A_{IN}(x) S \tag{2.16a}$$

$$|a\rangle_{OUT} = S^{\dagger} |a\rangle_{IN}, \tag{2.16b}$$

matrix elements of this operator are interpreted as the probability amplitude for a system originally in a state labeled by quantum numbers $\underline{a}$ at $t = -\infty$ to end up in a state labeled by quantum numbers $\underline{b}$ at $t = +\infty$, $S_{ba} \equiv {}_{OUT}\langle b|a\rangle_{IN} = {}_{IN}\langle b|S|a\rangle_{IN}$. In the Schwinger formalism (3.B), one constructs matrix elements of asymptotic particle states by functional differentiation of $\langle 0,t_1|0,t_2\rangle_{j\neq 0}$ with respect to (renormalized) sources of corresponding asymptotic arguments, subsequently taking the limit of zero source strength; and one finds, in effect, that $\lim \langle b,t_1|a,t_2\rangle\Big|_{\substack{t_1 \to +\infty \\ t_2 \to -\infty}}$ is equivalent to $_{OUT}\langle b|a\rangle_{IN}$ for arbitrary states $\underline{a},\underline{b}$.

Because the A_{IN} and A_{OUT} are to satisfy the same free field equations of motion, and are to have the same equal-time commutation relations, one infers the existence of a unitary transformation connecting them, as in (2.16a), an inference which could be proven were there but a finite number of degrees of freedom in the problem. We shall here assume (2.16) for both boson and fermion IN- and OUT- fields, and for the states constructed from them.

To demonstrate the connection between the S-matrix and the multipoint Green's functions obtained by functional differentiation of Z, it is simplest to consider first the case of a single scalar boson field, and its corresponding generating operator $T\{j\}$. From (2.3) there follows the relation

$$\frac{\delta T}{\delta j(x)} = iT_x^\infty A(x)T_{-\infty}^x, \tag{2.17}$$

and from (2.13)

$$\lim_{x_0 \to -\infty} [\frac{\delta T}{\delta j(x)} - iT\sqrt{Z_3}\, A_{IN}(x)] = 0, \tag{2.18a}$$

$$\lim_{x_0 \to +\infty} [\frac{\delta T}{\delta j(x)} - i\sqrt{Z_3}\, A_{OUT}(x)T] = 0, \tag{2.18b}$$

again to be understood as relations between matrix elements. Statements equivalent to (2.18) may be written in the form of (2.15),

$$\frac{\delta T}{\delta j(x)} = i\sqrt{Z_3}\, T A_{IN}(x) + \int \Delta_R(x-y) K_y \frac{\delta T}{\delta j(y)} \cdot d^4y, \quad (2.19a)$$

$$\frac{\delta T}{\delta j(x)} = i\sqrt{Z_3}\, A_{OUT}(x) T + \int \Delta_R(x-y) K_y \frac{\delta T}{\delta j(y)} \cdot d^4y, \quad (2.19b)$$

and the difference of this last pair yields the relation

$$\sqrt{Z_3}\,\left(A_{OUT}(x)T - TA_{IN}(x)\right) = i\int d^4y \cdot \Delta(x-y) K_y \frac{\delta T}{\delta j(y)} . \quad (2.20)$$

Multiplying both sides of (2.20) by S, and introducing (2.16a), one obtains the equation

$$[A_{IN}(x), ST] = iZ_3^{-\frac{1}{2}} \int d^4y \cdot \Delta(x-y) K_y \frac{\delta}{\delta j(y)} (ST), \quad (2.21)$$

which must be satisfied by the combination $ST\{j\}$.

At this stage it is useful to define the normal- or Wick-ordered exponential, $:e^{\int A_{IN} f}: \equiv e^{\int A_{IN}^{(-)} f}\, e^{\int A_{IN}^{(+)} f}$ where $\int A_{IN}^{(\pm)} f = \int d^4z A_{IN}^{(\pm)}(z) f(z)$. Since the commutator of $A_{IN}^{(+)}$ with $A_{IN}^{(-)}$ is a c-number, one may easily verify (Appendix D) the relation

$$[A_{IN}(x), :e^{\int A_{IN} f}:] = :e^{\int A_{IN} f}:\, i\int d^4y \Delta(x-y) f(y), \quad (2.22)$$

and a comparison with (2.21) then suggests the form[4]

$$ST = :e^{Z_3^{-\frac{1}{2}} \int A_{IN} \vec{K} \frac{\delta}{\delta j}}: F\{j\}. \quad (2.23)$$

It should perhaps be emphasized that the Klein-Gordon operator K_y in (2.23) is to act upon the coordinate freed by the functional differentiation operator

$\delta/\delta j(y)$, and <u>not</u> backwards upon A_{IN} (which would give just zero). Because of the special property $<:e^{\int A_{IN} f}:> = 1$, it follows that $F\{j\} = <ST\{j\}>$. Further, if there are no external fields present, the OUT-vacuum state, $_{OUT}<0| = {}_{IN}<0|S$, is physically indistinguishable from the IN-vacuum state, which means that the two can differ only by a phase, $_{OUT}<0| = e^{i\phi}{}_{IN}<0|$ with ϕ a real (and usually infinite) number. If, however, one is dealing with a situation in which it is convenient to introduce an external field A^{ext}, produced by an external current J^{ext} with Fourier components capable of creating particles (e.g., a radio antenna), then the OUT-vacuum will receive contributions from all the IN-states, and the vacuum expectation value of the S-matrix is not simply a phase factor. In this case it is appropriate to generalize the previous definition of Z to $<S>Z\{j,J^{ext}\} \equiv <ST>$ for this turns out to be the quantity directly calculable from the field equations and equal time commutation relations. In the absence of external fields, the phase factor is frequently omitted, since it contributes nothing to the calculation of cross sections; but it is just as easy to handle the general situation by writing

$$S = <S\{J^{ext}\}> :e^{Z_3^{-\frac{1}{2}} \int A_{IN} \vec{K} \frac{\delta}{\delta j}}: Z\{j,J^{ext}\}\Big|_{j=0}, \qquad (2.24)$$

where the nonexternal sources $j(x)$ are to be set equal to zero after the functional differentiation operations of (2.24) have been performed; with this generalized reduction formula one exhibits the S-matrix in terms of Wick-ordered A_{IN} fields and associated operations upon the generating functional. When fermions are included, the entire analysis goes through in a similar manner, with the result

$$\frac{S}{\langle S\rangle} = :e^{Z_3^{-\frac{1}{2}}\int A_{IN}\vec{K}\frac{\delta}{\delta j} + Z_2^{-\frac{1}{2}}\int[\bar{\psi}_{IN}\vec{\mathcal{D}}\frac{\delta}{\delta\bar{\eta}} - \frac{\delta}{\delta\eta}\overleftarrow{\bar{\mathcal{D}}}\psi_{IN}]} :Z\{j,\eta,\bar{\eta}\}\Big|_0 \tag{2.25}$$

with the artificial sources $j,\eta,\bar{\eta}$ set equal to zero after all the functional differentiations have been performed.

D. A Bremsstrahlung Example

In a perturbation expansion of the S-matrix, the $Z_{2,3}$ factors of (2.25) are absorbed into the definition of the renormalized charge, and do not appear in the probabilities for any physical process; this is briefly discussed in Chapter 6, after simple radiative corrections to the n-point Green's functions are described. A useful example of the way in which this reduction formula may be employed occurs in the process of multiple Bremstrahlung. The simplest such exercise is to calculate the probability amplitude for the emission of n photons when an electron is scattered by an external field; the resolution of the infrared difficulties of this canonical problem, while of fundamental importance to quantum electrodynamics, has provided the arithmetic structures appropriate to subsequent relativistic eikonal formulations. One requires the S-matrix element

$$\langle p',s';k_1,\varepsilon_{\mu_1}(k_1);\cdots k_n,\varepsilon_{\mu_n}(k_n)|S|p,s\rangle \tag{2.26}$$

where p,s and p',s' denote the initial and final electron momenta and spins, and the k_i with polarization indices ε_{μ_i} describe the emitted photons.

The states are represented by

$$|p,s\rangle = b_s^\dagger(p)\rangle \tag{2.27a}$$

and

$$\langle p',s';k_1,\varepsilon_1;\cdots k_n,\varepsilon_n| = \frac{1}{\sqrt{n!}}\,\varepsilon_{\mu_1}(k_1)\cdots\varepsilon_{\mu_n}(k_n)\langle b_{s'}(p')$$

$$\cdot\; a_{\mu_1}(k_1)\cdots a_{\mu_n}(k_n), \qquad (2.27b)$$

and these IN-field operators are to be commuted through the Wick-ordered brackets of (2.25). Each time this occurs, the factors

$$\int dz_i\,[a_\mu(k_i),A_\nu^{IN}(z_i)]\vec{K}_z\,\frac{\delta}{\delta j_\nu(z)} \qquad (2.28a)$$

and

$$\sum_{\alpha,\beta}\int d^4x'\,[b_{s'}(\vec{p}\,'),\bar{\psi}^{\alpha}_{IN}(x')]\,(\vec{\mathcal{D}}_{x'})_{\alpha\beta}\,\frac{\delta}{\delta\bar{\eta}_\beta(x')} \qquad (2.28b)$$

and

$$-\sum_{\alpha,\beta}\int d^4x\,\frac{\delta}{\delta\eta_\alpha(x)}\,(\overleftarrow{\bar{\mathcal{D}}}_x)_{\alpha\beta}[\psi^\beta_{IN}(x),b_s^\dagger(\vec{p})] \qquad (2.28c)$$

will appear, with the functional derivatives operating upon the generating functional. From (2.28) and the free-particle commutation and anticommutation relations, one easily obtains for (2.26)

$$-(2\pi)^{-\frac{3}{2}(n+2)}\,\frac{1}{\sqrt{n!}}\sum_{\mu_1}\varepsilon_{\mu_1}(k_1)\cdots\sum_{\mu_n}\varepsilon_{\mu_n}(k_n)\left[\frac{m}{EE'}\right]^{\frac{1}{2}}\cdot$$

$$\cdot\;[2\omega_1\cdots 2\omega_n]^{-\frac{1}{2}}\cdot\int d^4x'\,e^{-ip'x'}\,\bar{u}^{\alpha'}_{s'}(p')(\vec{\mathcal{D}}_{x'})_{\alpha'\beta'}\;\cdot$$

$$\cdot\;\int d^4x\,e^{ip\cdot x}(\vec{\mathcal{D}}_x)_{\alpha\beta}u^\beta_s(p)\cdot\langle S\rangle\cdot\int dz_1\cdots\int dz_n\,e^{-i(k_1z_1+\cdots+k_nz_n)}$$

$$\cdot \vec{K}_{z_1} \cdots \vec{K}_{z_n} \frac{\delta}{\delta j_{\mu_1}(z_1)} \cdots \frac{\delta}{\delta j_{\mu_n}(z_n)} \frac{\delta}{\delta \bar{\eta}_{\beta'}(x')} \frac{\delta}{\delta \eta_\alpha(x)} Z \Big|_o \tag{2.29}$$

From this one sees that the passage from generating functional to S-matrix element involves the calculation of $n + 2$ functional derivatives of Z, the operation upon this quantity with Klein-Gordon or Dirac operator, as appropriate, and finally, the Fourier transform of all configuration space dependence in terms of the appropriate mass shell momenta; these last two steps are frequently referred to as mass shell amputation. In order to proceed, one must next exhibit the generating functional in its dependence upon all the sources.

Notes

1. The discussion of this section follows that given by K. Symanzik, Lectures at UCLA, 1960.

2. J. Schwinger, Harvard Lectures (1954).

3. C. N. Yang and D. Feldman, Phys. Rev., 79, 972 (1950).

4. In this way, Wick's theorem (G. C. Wick, Phys. Rev., 80, 268 (1950)) becomes a simple statement concerning the normal ordering of an exponential functional operator.

Chapter 3

CONSTRUCTION OF THE GENERATING FUNCTIONAL

In this chapter, two different methods of constructing the generating functional will be given and extended to several interacting fields, with some care taken in the definition of products of fields appearing at the same space-time point.

A. The Symanzik Construction

The simplest situation deals with the self-interaction of a single, spinless, boson, scalar, Hermitian field $A(x)$. The problem is defined by variation of the Lagrangian density L, where $L = L_o + L'$, and

$$L_o = -\frac{1}{2}[\mu^2 A^2 + \sum_\mu (\partial_\mu A)^2], \tag{3.1}$$

$$L' = -\frac{g}{n!} A^n(x). \tag{3.2}$$

Here L_o denotes the canonical, noninteracting part of the Lagrangian, in terms of the unrenormalized boson mass μ, while L' denotes the interaction part of the Lagrangian, proportional to the unrenormalized charge, or coupling constant, g. For $n = 3$, one has the so-called Ward, Hurst, Thirring model, with structure resembling that of QED. For any odd $n \geq 3$, however, the particle states do not exhibit a lower bound to the Hamiltonian (Appendix E), and hence such theories cannot be taken seriously. Further, for $n \geq 5$ in four-dimensional space-time, such theories are nonrenormalizable. Products of field operators at the same point are notoriously singular objects, and proper definitions must be supplied in each case.

The field equation resulting from the variation of L is

$$K_x A(x) = -\frac{g}{(n-1)!} A^{n-1}(x), \tag{3.3}$$

which, together with the assumed ETCR (1.2), (1.3),

will be used to construct $Z\{j\}$. One has, from (2.1) and (2.3),

$$\frac{1}{i}\frac{\delta}{\delta j(x)} Z = \langle T_x^{\infty} \cdot A(x) \cdot T_{-\infty}^{x} \rangle, \tag{3.4}$$

and hence

$$\begin{aligned} K_x \cdot \frac{1}{i}\frac{\delta}{\delta j(x)} Z = &\langle T_x^{\infty} \cdot K_x A(x) \cdot T_{-\infty}^{x} \rangle \\ &+ \partial_0^2 \langle T_x^{\infty} \cdot A(x) \cdot T_{-\infty}^{x} \rangle \\ &- \langle T_x^{\infty} \cdot \partial_0^2 A(x) \cdot T_{-\infty}^{x} \rangle. \end{aligned} \tag{3.5}$$

The first term on the RHS of (3.5) is given by (3.3) as

$$-\frac{g}{(n-1)!} \langle T_x^{\infty} \cdot A^{n-1}(x) \cdot T_{-\infty}^{x} \rangle = -\frac{g}{(n-1)!} \left(\frac{1}{i}\frac{\delta}{\delta j(x)}\right)^{n-1} Z \tag{3.6}$$

while the remaining terms may be evaluated with the aid of the ETCR to give

$$\begin{aligned} &-i\int d^3x' j(\vec{x}',x_0) \langle T_x^{\infty} \cdot [A(\vec{x}',x_0),\partial_0 A(\vec{x},x_0)] \cdot T_{-\infty}^{x} \rangle \\ &= j(x) \langle T_x^{\infty} \cdot T_{-\infty}^{x} \rangle = j(x) Z. \end{aligned} \tag{3.7}$$

Combining (3.6) and (3.7) one obtains the functional differential equation

$$K_x \cdot \frac{1}{i}\frac{\delta}{\delta j(x)} Z = j(x) Z - \frac{g}{(n-1)!} \left(\frac{1}{i}\frac{\delta}{\delta j(x)}\right)^{n-1} Z. \tag{3.8}$$

In order to solve (3.8), consider first the special case of free-fields, $g = 0$; then (3.8) is simplified to

$$K_x \frac{1}{i} \frac{\delta}{\delta j(x)} Z^{(o)} = j(x) Z^{(o)}, \tag{3.9}$$

which form suggests the ansatz

$$Z^{(o)}\{j\} = \exp \Omega\{j\},$$

and the necessary side condition $\Omega\{0\} = 0$. This form substituted into (3.9) yields the differential equation

$$K_x \frac{1}{i} \frac{\delta}{\delta j(x)} \Omega = j(x),$$

with solution

$$\frac{\delta}{\delta j(x)} \Omega = i \int \Delta_c(x-x')j(x')d^4x'. \tag{3.10}$$

In (3.10) the causal Green's function has been written as the appropriate solution of the inhomogeneous K-G equation, because Z is defined in terms of time ordered operators,

$$\frac{\delta}{\delta j(x)} \frac{\delta}{\delta j(y)} Z\Big|_0 = i^2 \langle (A(x)A(y))_+ \rangle .$$

It is not difficult to show that translational invariance (here the statement that $\langle A(x)\rangle = \langle A(0)\rangle$) rules out, for massive fields, the addition of an arbitrary solution of the homogeneous K-G equation to the RHS of (3.10); and hence the free-field generating functional is given by

$$Z^{(o)}\{j\} = \exp \frac{i}{2} \iint d^4x d^4y j(x) \Delta_c(x-y) j(y), \tag{3.11}$$

a form which will be abbreviated in all of the following by $\exp[\frac{i}{2} \int j\Delta_c j]$. Note that the factor of 1/2 arises because $\Delta_c(z)$ is an even function of its argument.

Incidentally, (3.11) and the zero-external field asymptotic condition (2.24), which is trivially true for free fields, permit one to derive an occasionally useful statement relating the time-ordered functional $T^{(o)}\{j\}$ to the related Wick-ordered quantity. Writing (2.23) with $Z = 1$ and $S = 1$ for the free field case, and inserting (3.11), one has

$$T^{(o)}\{j\} = :e^{\int A_{IN} K \frac{\delta}{\delta j}}:e^{\frac{i}{2}\int j \Delta_c j} \tag{3.12}$$

which is easily evaluated to give

$$\left(e^{i\int jA_{IN}}\right)_+ = :e^{i\int jA_{IN}}:e^{\frac{i}{2}\int j\Delta_c j}, \tag{3.13}$$

and should be compared with a result of Appendix D,

$$e^{i\int jA_{IN}} = :e^{i\int jA_{IN}}:e^{-\frac{1}{4}\int j\Delta_{(1)} j}. \tag{3.14}$$

A solution to (3.8) may be obtained from the ansatz

$$Z\{j\} = N^{-1} \cdot \exp F\left\{\frac{1}{i}\frac{\delta}{\delta j}\right\} \cdot Z^{(o)}\{j\}, \tag{3.15}$$

where N is a normalization constant and F a functional to be determined. Substitution of (3.15) into (3.8) yields

$$\begin{aligned} K_x \frac{1}{i}\frac{\delta}{\delta j(x)} Z &= N^{-1} \cdot \exp F\left\{\frac{1}{i}\frac{\delta}{\delta j}\right\} \cdot j(x)Z^{(o)} \\ &= j(x)Z + N^{-1}\left[\exp F\left\{\frac{1}{i}\frac{\delta}{\delta j}\right\}, j(x)\right]Z^{(o)}, \end{aligned} \tag{3.16}$$

and F must be chosen such that the last term on the RHS of (3.16) is equivalent to

$-\frac{g}{(n-1)!}\left(\frac{1}{i}\frac{\delta}{\delta j(x)}\right)^{n-1} Z$, which means

$$[\exp F\left\{\frac{1}{i}\frac{\delta}{\delta j}\right\}, j(x)] = -\frac{g}{(n-1)!}\left(\frac{1}{i}\frac{\delta}{\delta j(x)}\right)^{n-1} \cdot$$

$$\cdot \exp F\left\{\frac{1}{i}\frac{\delta}{\delta j}\right\}.$$

A glance at the forms of Appendix D suggests the solution,

$$F\left\{\frac{1}{i}\frac{\delta}{\delta j}\right\} = -i\,\frac{g}{n!}\int d^4z\left(\frac{1}{i}\frac{\delta}{\delta j(z)}\right)^n,$$

and

$$Z\{j\} = N^{-1}\exp[-\frac{ig}{n!}\int\left(\frac{1}{i}\frac{\delta}{\delta j}\right)^n] \cdot \exp\frac{i}{2}\int j\Delta_c j, \qquad (3.17)$$

with the constant N determined by the condition $Z\{0\} = 1$. It should be noted that $F\{A\} = i\int L'\{A\}$ is simply the interaction part of the action operator, $W'\{A\}$. In fact, this solution was first obtained by Schwinger[1] directly from a suitable action principle, and immediately afterwards derived in the above form by Symanzik.[2] An elegant derivation, together with the formulation of a modified perturbation theory, has been given by Fradkin.[3]

B. The Schwinger Construction

It will be most useful to describe, in at least a cursory way, the corresponding derivation given by Schwinger, and for this purpose it is necessary to state the rudiments of his quantum mechanical action principle relevant to this field theoretic situation. From states labeled by complete sets of quantum numbers a' at some time t, one constructs the so-called transformation functions $\langle a',t|b',t\rangle$ which permit a change in the description of the system from the set of a' variables to an equivalent set of b' variables. More generally, if the times are different, one has the probability amplitude $\langle a',t_1|b',t_2\rangle$ for the system to be in a state labeled by the a' at time t_1 if it was known to be in the state labeled by the b' at t_2, where $t_1 > t_2$. These quantities are assumed to obey the relations

$$\langle a',t_1|b',t_2\rangle^* = \langle b',t_2|a',t_1\rangle \tag{3.18a}$$

and

$$\sum_n \langle a',t_1|n',t\rangle\langle n',t|b',t_2\rangle = \langle a',t_1|b',t_2\rangle, \tag{3.18b}$$

the reality and closure properties, respectively. In (b), the summation is over a complete set of states defined at an arbitrary time t. The states $|a,t\rangle$ are defined as eigenstates of Hermitian operators $A(t)$ in the Heisenberg picture, and are not necessarily the time-dependent vectors of the Schroedinger picture. For the particle situation, however, and in the absence of external, time-dependent sources, $A(t)$ may be identified with the complete set of commuting Heisenberg operators possessing constant energy, charge, ... eigenvalues, and the $|a,t\rangle$ become conventional Schroedinger states.

The action principle provides differential statements concerning the infinitesimal variations of these functions, from which immediately follow their explicit, integral solutions. The essential assumption is that any variation $\delta\langle a',t_1|b',t_2\rangle$ may be represented as the matrix element of an infinitesimal hermitian operator between the same states,

$$\delta\langle a',t_1|b',t_2\rangle = i\langle a',t_1|\delta W_{12}|b',t_2\rangle \tag{3.19}$$

where δW_{12} denotes the appropriate variation of the action operator $W_{12} = \int_{t_2}^{t_1} d^4x L$. With the forms (3.1), (3.2), and under the simplest variation $A \to A + \delta A$, with δA a c-number, the Euler equation (3.3) is obtained by requiring a zero value of (3.19), if the variations δA are to vanish at the end-point times.

More generally, if such end-point variations are permitted, (3.19) will be given in terms of the matrix elements of operators defined at the end-point times only. These operators are called generators because they define infinitesimal unitary transformations corresponding to Lorentz translations and rotations, and also transformations corresponding to possible changes of the quantum mechanical coordinates of the system; from these latter generators, one may obtain the fundamental ETCR (1.3), which have here been separately assumed in the previous discussion.

However, (3.19) is intended to hold for more general types of variations; in particular, consider the Lagrangian $L = L_0\{A\} + L'\{A\} + jA$ where $j(x)$ is an arbitrary, c-number source function. The states of this system now depend upon this "external" current source. The simplest quantity to consider is the probability amplitude that the vacuum at time t_2 remains a vacuum at time t_1 under the influence of the source which could, in principle, pump enough energy and momentum into the system to create real particles. Under the infinitesimal change $j \to j + \delta j$ there is a corresponding change in this amplitude, given by (3.19). The fields A and $\partial_\mu A$ appearing in W are themselves functionals of j, and variations of W due to the induced variations of A and $\partial_\mu A$ must vanish by virtue of the Euler equation (of the form (3.3) with an extra term $j(x)$ on the RHS); hence, it is sufficient to consider just the variation of the explicit j dependence of W, so that (3.19) simply yields

$$\frac{1}{i}\frac{\delta}{\delta j(x)}\langle 0,t_1|0,t_2\rangle = \langle 0,t_1|A(x)|0,t_2\rangle \qquad (3.20)$$

if $t_1 \geq x_0 \geq t_2$, and zero otherwise.

Repeated functional differentiation may be evaluated in a similar way, with the aid of (3.18). For example, if $t_1 > x_0 > y_0 > t_2$,

$$\left(\frac{1}{i}\right)^2 \frac{\delta}{\delta j(y)} \frac{\delta}{\delta j(x)} <0,t_1|0,t_2>$$

$$= \frac{1}{i} \frac{\delta}{\delta j(y)} \sum_n <0,t_1|A(x)|n',x_o><n',x_o|0,t_2>$$

$$= \sum_n <0,t_1|A(x)|n',x_o><n',x_o|A(y)|0,t_2> \tag{3.21}$$

$$= <0,t_1|A(x)A(y)|0,t_2>.$$

On the other hand, if $t_1 > y_o > x_o > t_2$, the same technique produces (3.21) with the operators interchanged; and hence the complete statement is

$$\left(\frac{1}{i}\right)^2 \frac{\delta}{\delta j(x)} \frac{\delta}{\delta j(y)} <0,t_1|0,t_2> = <0,t_1|\left(A(x)A(y)\right)_+|0,t_2>$$

$$t_1 > x_o, y_o > t_2, \tag{3.22}$$

or in general,

$$\left(\frac{1}{i}\right)^n \frac{\delta}{\delta j(x_1)} \cdots \frac{\delta}{\delta j(x_n)} <0,t_1|0,t_2> =$$

$$<0,t_1|\left(A(x_1) \cdots A(x_n)\right)_+|0,t_2>,$$

$$t_1 > x_{1,0} \cdots x_{n,0} > t_2. \tag{3.23}$$

Notice that (3.23) provides a definition, in part, for the matrix elements of a product of operators taken at the same point, in the sense that such limits are to be understood in a symmetric way.

Equation (3.23) may be used to provide a connection between the generating functional of (3.17) and the probability amplitude $\langle 0,t_1|0,t_2\rangle$ in the limits $t_1 \to +\infty$, $t_2 \to -\infty$, since the RHS of (3.23) generates the coefficient of that term in the Taylor expansion of $\langle 0,+\infty|0,-\infty\rangle$ proportional to n factors of j,

$$\left(\frac{1}{i}\right)^n \left[\frac{\delta}{\delta j(x_1)} \cdots \frac{\delta}{\delta j(x_n)} \langle 0,+\infty|0,-\infty\rangle\right]_{j=0}$$

$$= \langle 0,+\infty|\big(A(x_1) \cdots A(x_n)\big)_+|0,-\infty\rangle\Big|_{j=0} \tag{3.24}$$

where the states and operators on the RHS of (3.24) are the appropriate quantities in the absence of the external field j. From (2.14) and (2.16b) these yield a combination equivalent to

$${}_{IN}\langle 0|S\big(A(x_1) \cdots A(x_n)\big)_+|0\rangle_{IN}$$

$$= \frac{1}{i^n}\frac{\delta}{\delta j(x_1)} \cdots \frac{\delta}{\delta j(x_n)} \langle ST\{j\}\rangle\Big|_{j=0}, \tag{3.25}$$

and hence

$$\langle 0,+\infty|0,-\infty\rangle_{j\neq 0} = \langle S\rangle Z\{j\}. \tag{3.26}$$

There may be other external fields J in the general problem, such that both sides of (3.26) depend upon J in the form

$$\langle 0,+\infty|0,-\infty\rangle_{j,J} = \langle S\{J\}\rangle Z\{j,J\}$$

$$= \langle S\{J\} \cdot \left(\exp i{\textstyle\int} jA\right)_+\rangle, \tag{3.27}$$

where $Z\{0,J\} = 1$. The essential point is that while J and j both appear as "external" fields in the Schwinger amplitude, all the j dependence has been explicitly written in the ordered exponent of the Symanzik amplitude, on the RHS of (3.27); the operator which appears in the latter may be an implicit functional of J, but not of j.

The action principle may be directly used to construct the Schwinger amplitude. Again, one first considers the free-field case corresponding to $g = 0$ to obtain

$$\frac{1}{i}\frac{\delta}{\delta j(x)} \langle 0,t_1|0,t_2\rangle_{g=0} = \langle 0,t_1|A(x)|0,t_2\rangle_{g=0}, \quad (3.28)$$

and hence

$$K_x \frac{1}{i}\frac{\delta}{\delta j(x)} \langle 0,t_1|0,t_2\rangle_{g=0} = j(x)\langle 0,t_1|0,t_2\rangle_{g=0}, \quad (3.29)$$

since A satisfies the simple relation $K_x A(x) = j(x)$. Exactly the same arguments used in the construction of $Z^{(o)}$ are relevant here, and one obtains

$$\langle 0,t_1|0,t_2\rangle_{g=0} = \exp[\frac{i}{2}\int_{t_2}^{t_1} j\Delta_c j], \quad (3.30)$$

where both time integrands of the exponent of (3.30) are understood to be limited to the range $t_1 \geq t \geq t_2$.

Application of the action principle to variations of the coupling parameter, $g \to g + \delta g$, yields

$$\frac{\delta}{\delta g} \langle 0,t_1|0,t_2\rangle = -\frac{i}{n!}\langle 0,t_1|\int_{t_2}^{t_1} d^4x A^n(x)|0,t_2\rangle$$

$$= -\frac{i}{n!}\int_{t_2}^{t_1} d^4x \left(\frac{1}{i}\frac{\delta}{\delta j(x)}\right)^n \cdot \langle 0,t_1|0,t_2\rangle, \quad (3.31)$$

where one again considers only the explicit dependence of W on g. The solution of (3.31) is elementary,

$$\langle 0,t_1|0,t_2\rangle = \exp[-\frac{i}{n!} g\int\left(\frac{1}{i}\frac{\delta}{\delta j}\right)^n] \cdot \langle 0,t_1|0,t_2\rangle_{g=0}, \tag{3.32}$$

and in the limits $t_1 \to \infty$, $t_2 \to -\infty$, one sees the correspondence of (3.32) to (3.17); the interrelation of these functionals, according to (3.26), shows that the normalization constant N of (3.17) may be identified as the vacuum-to-vacuum amplitude $\langle S\rangle$. The same technique, in any field theory with canonical L_0 and causal interaction $L'\{A\}$ always produces a generating functional of the form

$$\langle S\rangle Z\{j\} = \exp[iW'\left\{\frac{1}{i}\frac{\delta}{\delta j}\right\}] \cdot \exp[\frac{i}{2}\int j\Delta_c j]. \tag{3.33}$$

Theories with noncanonical L_0, such as those defined by the chiral invariant, nonpolynomial Lagrangians, require a separate discussion, as in Chapter 4.

C. Several Interacting Fields

Generalizations of (3.33) to more realistic theories are immediate. For the case of two boson fields, ϕ and A, defined by the Lagrangian

$$L = L_0\{A\} + L_0\{\phi\} - \frac{g}{2}\phi^2 A,$$

one writes the generating functional

$$Z\{j,k\} = \langle(\exp i\int[jA + k\phi])_+\rangle \tag{3.34}$$

and may carry through either the Schwinger or Symanzik construction to obtain the free-field functional ($g = 0$),

$$Z^{(0)}\{j,k\} = \exp[\frac{i}{2}\int j\Delta_c^{(\mu)} j + \frac{i}{2}\int k\Delta_c^{(M)} k] \tag{3.35}$$

and the representation of (3.34)

$$Z\{j,k\} = N^{-1} \cdot \exp[-i\,\frac{g}{2}\int\left(\frac{1}{i}\frac{\delta}{\delta k}\right)^2\left(\frac{1}{i}\frac{\delta}{\delta j}\right)] \cdot Z^{(o)}\{j,k\}. \tag{3.36}$$

In (3.35), $\Delta_c^{(\mu)}$ and $\Delta_c^{(M)}$ denote the free, causal propagators of the fields A and ϕ, of unrenormalized mass μ and M, respectively; again, $N = \langle S\rangle$ and is a pure phase factor. The Feynman graphs frequently drawn to represent the interaction between the quanta of these fields are interpreted in terms of the production of singly emitted A-particles by a ϕ-particle, or of the conversion of single A-particles into pairs of ϕ-particles.

There is an interesting limiting situation, arising in the bremstrahlung of low energy photons by a heavy, or fast moving charged particle, which may be described by replacing the $-(g/2)\phi^2$ term of the previous interaction Lagrangian by a real, c-number function $J(x)$. This represents an essentially classical-current description of the ϕ-particles, which are treated as unaffected by their emission of A-quanta. Such neglect of radiation reaction is only an approximation, but a very useful one; in this model, the ϕ-particles only exist to provide the prescribed c-number current $J(x)$, and all other ϕ-dependence may be dropped from the Lagrangian. Thus, one considers $L = L_0\{A\} + JA$, with the generating functional defined by

$$\langle S\{J\}\rangle Z\{j;J\} = \langle S\{J\}(\exp i\textstyle\int jA)_+\rangle. \tag{3.37}$$

Either form of construction yields

$$\langle S\rangle Z = \exp[i\int J\left(\frac{1}{i}\frac{\delta}{\delta j}\right)] \cdot \exp[\frac{i}{2}\int j\Delta_c j]. \tag{3.38}$$

The operator $\exp[\int J\,\frac{\delta}{\delta j}]$ of (3.38) expresses a simple translation of the j dependence of any functional $F\{j\}$ upon which it operates into $F\{j+J\}$, and hence

$$\langle S\rangle Z = \exp[\frac{i}{2}\int (J+j)\Delta_c(J+j)]. \tag{3.39}$$

With the normalization condition $Z\{0;J\} = 1$, one obtains

$$\langle S\{J\}\rangle = \exp[\frac{i}{2}\int J\Delta_c J], \tag{3.40}$$

a result which could have been read off directly from (3.30). The S-matrix is given by the application of the general reduction formula (2.24), with $Z_3 = 1$ because there are no quantized fluctuations of the A field in this model situation. Using the same analysis as that which led from (3.12) to (3.13), one easily finds

$$S = \langle S\{J\}\rangle :\exp[i\int JA_{IN}]:. \tag{3.41}$$

The exponent of (3.40) is no longer purely imaginary; with

$$\Delta_c(x) = \frac{i}{2}\Delta_{(1)}(x) - \frac{1}{2}\varepsilon(x_o)\Delta(x) \quad \text{and}$$

$$\frac{i}{2}\int J\Delta_c J \equiv -\frac{1}{2}W + i\xi,$$

where W and ξ are real, one has

$$W = \frac{1}{2}\iint d^4xd^4yJ(x)\Delta_{(1)}(x-y)J(y) \tag{3.42a}$$

and

$$\xi = -\frac{1}{4}\iint d^4xd^4yJ(x)\varepsilon(x_o-y_o)\Delta(x-y)J(y). \tag{3.42b}$$

Inserting a Fourier representation of the real currents $J(z)$, (3.42a) can be written as

$$W = \frac{1}{2}\int d^4k\delta(k^2 + \mu^2)|\tilde{J}(k)|^2, \tag{3.43}$$

indicating that W will be real and positive for those external currents whose frequency dependence can excite mass-shell quanta of the A field. When this is the case, the probability of the vacuum remaining a vacuum over an arbitrarily large time interval, $P_o = |\langle S\rangle|^2 = \exp[-W]$, must be less than unity.

More generally, the probability of the vacuum to turn into a state of n particles can also be obtained in this simple model.[4] It is

$$P_n = \sum_{\gamma_n} |{}_{IN}\langle n,\gamma_n|S|0\rangle_{IN}|^2 \tag{3.44}$$

where the γ_n label on the state of n particles refers to all of the n-particle phase space variables which must be summed over, taking into account energy and momentum conservation; with this notation, the closure relation would read $\sum_n \sum_{\gamma_n} |n,\gamma_n\rangle\langle n,\gamma_n| = 1$. From (3.41) one has

$$\langle n,\gamma_n|S\rangle = \frac{1}{n!}\langle n,\gamma_n|\left(i\int JA_{IN}^{(-)}\right)^n\rangle \cdot \langle S\rangle, \tag{3.45}$$

where only that term of the exponential's expansion which contains n factors of $A_{IN}^{(-)}$ can connect the IN vacuum to an IN-state of n particles. Substituting (3.45) into (3.44),

$$P_n = e^{-W}\left(\frac{1}{n!}\right)^2 \sum_{\gamma_n} \langle\left(\int JA_{IN}^{(+)}\right)^n|n,\gamma_n\rangle\langle n,\gamma_n|\left(\int JA_{IN}^{(-)}\right)^n\rangle, \tag{3.46}$$

and one notes that the sum over the intermediate states for fixed n may be extended to range over all n,

$$\sum_{\gamma_n} |n,\gamma_n\rangle\langle n,\gamma_n| \to \sum_{n',\gamma_{n'}} |n',\gamma_{n'}\rangle\langle n',\gamma_{n'}|, \tag{3.47}$$

if only the original n index is maintained for the number of $A^{(\pm)}_{IN}$ factors; this is true because the only contribution of $\langle n',\gamma_{n'}|(\int JA^{(-)}_{IN})^n\rangle$ occurs when $n' = n$. But the RHS of (3.47) is unity by the closure relation, and hence

$$P_n = e^{-W}\left(\frac{1}{n!}\right)^2 \langle (\int JA^{(+)}_{IN})^n (\int JA^{(-)}_{IN})^n \rangle .$$

This vacuum expectation value has the form $\langle a^n b^n \rangle$, where $[a,b]$ = c-number, $a|0\rangle_{IN} = 0$ and ${}_{IN}\langle 0|b = 0$. Under these conditions it immediately follows from the forms of Appendix D that $\langle a^n b^n\rangle = n!([a,b])^n$. With the aid of (1.5) one then obtains $[a,b] = i\int J\Delta^{(+)}J$, which by symmetry may also be written as $\frac{1}{2}\int J\Delta_{(1)}J$. Recalling (3.42a), one obtains the familiar Poisson form

$$P_n = \frac{W^n}{n!}\, e^{-W} \tag{3.48}$$

as the probability distribution for the emission of n particles by the external source. The normalization condition, $\sum_n P_n = 1$ is of course consistent with this interpretation. It should be noted that W represents the average multiplicity, $W = \sum_n nP_n$.

A theory of fermions coupled to an external, c-number boson field is easily handled in a similar way. The Lagrangian here is $L = L_o(\psi,\bar{\psi}) + L'$, with

$$L_o = -\bar{\psi}(m + \gamma\cdot\partial)\psi, \tag{3.49a}$$

and

$$L' = g\bar{\psi}\psi A^{ext}, \tag{3.49b}$$

while the generating functional

$$\langle S\{A^{ext}\}\rangle Z\{\eta,\bar{\eta};A^{ext}\} = \langle S\{A^{ext}\}(\exp i\int[\bar{\psi}\eta + \bar{\eta}\psi])_+\rangle$$

is rapidly determined to be

$$\langle S\rangle Z = \exp(ig\int[-\frac{1}{i}\frac{\delta}{\delta\eta}]A^{ext}[\frac{1}{i}\frac{\delta}{\delta\bar{\eta}}]) \cdot Z^{(o)}\{\eta,\bar{\eta}\}. \quad (3.50)$$

The free-field functional is given by

$$Z^{(o)} = \exp[i\iint d^4x d^4y \sum_{\alpha,\beta} \bar{\eta}_\alpha(x) S_c^{\alpha\beta}(x-y)\eta_\beta(y)], \quad (3.51)$$

a form which will subsequently be abbreviated as $\exp[i\int\bar{\eta}S_c\eta]$. The steps leading to these equations are completely analogous to those followed for the boson case. If the interaction is that of an external electromagnetic field interacting with the fermion current, the only overt change in these formulae is the inclusion of a four-vector index,

$$L' = ie\sum_\mu \bar{\psi}\gamma_\mu\psi \cdot A_\mu^{ext} \quad (3.52)$$

and

$$\langle S\rangle Z = \exp\left(-e\int \frac{\delta}{\delta\eta}\gamma \cdot A^{ext}\frac{\delta}{\delta\bar{\eta}}\right) \cdot \exp(i\int\bar{\eta}S_c\eta). \quad (3.53)$$

Finally, in a coupled theory of fermions and bosons, here illustrated by QED, one writes a Lagrangian of form $L = L_o + L'$ where L_o denotes the sum of (3.49a) and (3.1), the latter with the replacement $A \to A_\mu$ and the mass $\mu \to 0$; for L' one adopts the familiar classical form, rewritten in terms of operators only,

$$L' = ie\bar{\psi}\gamma \cdot A\psi. \quad (3.54)$$

The generating functional

$$Z\{j_\mu,\eta,\bar{\eta}\} = \langle(\exp i\int[j_\nu A_\nu + \bar{\eta}\psi + \bar{\psi}\eta])_+\rangle$$

may then be immediately written in the form

$$\langle S\rangle Z = \exp[ie\int \frac{\delta}{\delta\eta}\left(\gamma \cdot \frac{\delta}{\delta j}\right)\frac{\delta}{\delta\bar{\eta}}] \cdot$$

$$\cdot \exp[\frac{i}{2}\int j_\mu D_c^{\mu\nu} j_\nu + i\int\bar{\eta}S_c\eta], \tag{3.55}$$

where $D_c^{\mu\nu} = \delta_{\mu\nu}\Delta_c\Big|_{\mu^2=0}$. Equation (3.55) will be generalized to arbitrary relativistic gauges in Chapter 5.

D. Rearrangements/Grouping of Feynman Graphs

Each of these specific forms for the generating functional is exact, and is complete in the sense that expansion in powers of the appropriate coupling constant produces all the (unrenormalized) perturbative contributions to every n-point Green's function, and thence to every S-matrix element. In each coupled-field case there exists a simple rearrangement which when performed leads to an advantageous grouping of terms into distinct classes of Feynman graphs, in the sense of being able to distinguish certain closed-loop insertions into an arbitrary graph. It is also a useful rearrangement for exhibiting the relationship between relativistic, quantum field theory (2nd quantization) and relativistic potential theory (1st quantization).

A simple, algebraic derivation of the formula

$$\exp[-\frac{i}{2}\int\frac{\delta}{\delta j}A\frac{\delta}{\delta j}] \cdot \exp[\frac{i}{2}\int jBj] = \exp[\frac{i}{2}j\bar{B}j] \cdot \exp L \tag{3.56}$$

where

$$\bar{B} = B(1 - AB)^{-1} \quad \text{and} \quad L = \frac{1}{2}\,\mathrm{Tr}\,\ln(1 - AB)^{-1}$$

is sketched in Appendix F. Here, $A(x,y)$ and $B(x,y)$ are each symmetric functions of their arguments, while $\overline{B}(x,y) = \langle x|\overline{B}|y\rangle$ denotes that solution of the integral equation

$$\overline{B}_\lambda(x,y) = B(x,y) + \lambda\int d^4u\int d^4v B(x,u)A(u,v)\overline{B}_\lambda(v,y) \quad (3.57)$$

with $\lambda = 1$. The quantity L is defined by the parametric integral

$$L = \frac{1}{2}\int_0^1 d\lambda\int d^4x\int d^4y A(x,y)\overline{B}_\lambda(y,x) \quad (3.58)$$

and, as indicated in Appendix G, is a useful representation for the logarithm of the determinant of the matrix $(1-AB)^{-\frac{1}{2}}$, in the limit of continuous (coordinate) indices. An elementary generalization of (3.56) is

$$\exp[-\frac{i}{2}\int\frac{\delta}{\delta j}A\frac{\delta}{\delta j}]\cdot\exp[\frac{i}{2}\int jBj + i\int fj]$$

$$= \exp[\frac{i}{2}\int j\overline{B}j + i\int j(1-BA)^{-1}f + \frac{1}{2}\text{Tr}\ell n(1-BA)^{-1}$$

$$+ \frac{i}{2}\int fA(1-BA)^{-1}f], \quad (3.59)$$

and is frequently useful, as in Chapter 7.

With the aid of (3.56), the generating functional of (3.36) may now be rewritten in the form

$$NZ\{j,k\} = \exp[-\frac{i}{2}\int\frac{\delta}{\delta k}(ig\frac{\delta}{\delta j})\frac{\delta}{\delta k}]\cdot\exp[\frac{i}{2}\int k\Delta_c^{(M)}k]$$

$$\cdot\exp[\frac{i}{2}\int j\Delta_c^{(\mu)}j]$$

$$= \exp[\frac{i}{2}\int k\Delta_c^{(M)}(1-ig\frac{\delta}{\delta j}\cdot\Delta_c^{(M)})^{-1}k +$$

$$+ \frac{1}{2} \text{Tr}\ell n\left(1-ig \frac{\delta}{\delta j} \cdot\Delta_c^{(M)}\right)^{-1}] \cdot \exp[\frac{i}{2} \int j\Delta_c^{(\mu)} j],$$
(3.60)

where $\langle x|\frac{\delta}{\delta j}|y\rangle = \delta(x-y) \frac{\delta}{\delta j(x)}$. A slightly more convenient form results if a further rearrangement is made, based upon the relation of Appendix H,

$$F\{\frac{1}{i} \frac{\delta}{\delta j}\}\exp[\frac{i}{2} \int j\Delta_c j] = \exp[\frac{i}{2} \int j\Delta_c j]$$

$$\cdot \exp[- \frac{i}{2} \int \frac{\delta}{\delta A} \Delta_c \frac{\delta}{\delta A}]F\{A\}, \quad (3.61)$$

where $A(x) \equiv \int d^4y\Delta_c(x-y)j(y)$. In place of (3.60), one may then write

$$NZ\{j,k\} = \exp[\frac{i}{2} \int j\Delta_c^{(\mu)} j] \cdot \exp[- \frac{i}{2} \int \frac{\delta}{\delta A} \Delta_c^{(\mu)} \frac{\delta}{\delta A}]$$

$$\cdot \exp[\frac{i}{2} \int kG[A]k + L[A]], \quad (3.62)$$

where $\langle x|G[A]|y\rangle = G_c(x,y|gA)$ denotes the (relativistic, potential theory) propagator of the ϕ-field in the presence of the c-number source $A(z)$,

$$G_c(x,y|gA) = \langle x|\Delta_c^{(M)}\left(1+gA\Delta_c^{(M)}\right)^{-1}|y\rangle \quad (3.63)$$

and

$$L[A] = \frac{1}{2} \int_0^g dg' \int d^4xA(x)G_c(x,x|g'A). \quad (3.64)$$

The forms (3.60) or the equivalent (3.62) are quite useful in the practical application of the functional techniques to estimates of different physical processes. The presence of the "closed loop functional" $L[A]$ in these formulae is the sign that closed ϕ-loops are inserted in all possible ways, quite automatically by the formalism, in any scattering of A- or ϕ-particles; the frequent approximation which neglects all closed ϕ-loops is simply made by dropping this term, $L[A] \to 0$, and making the corresponding simplification dictated by normalization, $N \to 1$. Perhaps the simplest example arises in the elastic scattering of a pair of ϕ-particles, by the mechanism of exchange of all possible virtual A-particles. The four-point Green's function of interest is then

$$<\left(\phi(x_1)\phi(x_2)\phi(x_3)\phi(x_4)\right)_+>$$

$$= \frac{\delta}{\delta k(x_1)} \frac{\delta}{\delta k(x_2)} \frac{\delta}{\delta k(x_3)} \frac{\delta}{\delta k(x_4)} Z\Big|_{j=k=0}, \qquad (3.65)$$

which, from (3.62) in the absence of closed loops, simply becomes $M(x_1,x_2;x_3,x_4) + M(x_1,x_3;x_2,x_4) + M(x_1,x_4;x_2,x_3)$, where

$$M(x_1,x_2;x_3,x_4) = \exp[-\frac{i}{2}\int \frac{\delta}{\delta A} \Delta_c^{(\mu)} \frac{\delta}{\delta A}]$$

$$\cdot\ G(x_1,x_2|gA)G(x_3,x_4|gA)\Big|_{A=0}. \qquad (3.66)$$

The sum over the three permutations of M is required by the indistinguishability of the ϕ-bosons; were the ϕ-particles entering into the scattering process actually distinguishable, then (3.66) alone would be sufficient to describe the scattering of ϕ_I, which "enters" and "leaves" the scattering region with coordinates x_2 and x_1, respectively; and of ϕ_{II} with analogous coordinates x_4 and x_3. For clarity and simplicity, this situation is now assumed, and the I, II labels will be appended to the $G[gA]$ of (3.66).

A further rearrangement can be made, useful in the context of this and related (eikonal) problems, which is based upon an elementary formula of Appendix I,

$$\exp[-\frac{i}{2}\int \frac{\delta}{\delta A}\Delta_c \frac{\delta}{\delta A}]G_I[A]G_{II}[A]$$

$$= \exp[-i\int \frac{\delta}{\delta A_1}\Delta_c \frac{\delta}{\delta A_2}] \cdot \left(\exp[-\frac{i}{2}\int \frac{\delta}{\delta A_1}\Delta_c \frac{\delta}{\delta A_1}]G_I[A_1]\right)$$

$$\cdot \left(\exp[-\frac{i}{2}\int \frac{\delta}{\delta A_2}\Delta_c \frac{\delta}{\delta A_2}]G_{II}[A_2]\right)\Big|_{A_1=A_2=A}. \qquad (3.67)$$

In pictorial terms, the effect of the "linkage" operator on the LHS of (3.67) may be expressed in terms of the distinct roles played by the operators on the RHS. The expansion of each RHS exponential operator produces all possible Feynman graphs containing virtual A lines (or propagators) which begin and end on the same $G[A]$ as well as all virtual A lines which link G_I with G_{II}. The simplest sort of approximation is obtained by the neglect of all "self-linkages",

$$M(x_1,x_2;x_3,x_4) = \exp[-i\int \frac{\delta}{\delta A_1}\Delta_c^{(\mu)} \frac{\delta}{\delta A_2}] \cdot G_I(x_1x_2|gA_1)$$

$$\cdot G_{II}(x_3,x_4|gA_2)\Big|_{A_1=A_2=0}, \qquad (3.68)$$

and corresponds to the sum of all "ladder" and "crossed" graphs representing all the virtual A mesons exchanged between ϕ_I and ϕ_{II}. Frequently, in the literature, more effort has been expended to sum the set of ladder graphs (thereby selecting, in order g^{2n}, just one out of $n!$ possible terms) than is required to estimate the entire contribution of (3.68).

The form of (3.56) useful for fermion problems is

$$\exp[-i\int \frac{\delta}{\delta\eta} A \frac{\delta}{\delta\bar{\eta}}] \cdot \exp[i\int \bar{\eta}B\eta] = \exp[i\int \bar{\eta}\bar{B}\eta] \cdot \exp L,$$

$$\bar{B} = B(1+AB)^{-1}, \quad L = -\mathrm{Tr}\ \ell n(1+AB)^{-1}, \qquad (3.69)$$

where the differences of sign in the definitions of $\bar{B}$ and L relative to (3.56) should be noted; these are due to the anticommuting nature of the fermion sources which reflect the necessary Fermi-Dirac statistics of spin 1/2 particles. In particular, the origin of the Feynman graph rule "an extra minus sign for each closed fermion loop" can be seen in the change of sign of this L of (3.69), compared to the boson L of (3.56). Factors of 1/2 are missing here as well, corresponding to the lack of symmetry of the arguments of $A_{\alpha\beta}(x,y)$ and $B_{\alpha\beta}(x,y)$, while in the definition of L the trace summation includes a sum over Dirac indices.

With the aid of (3.69) the generating functional of (3.53) may be rewritten as

$$\langle S\rangle Z = \exp\left\{i\int \bar{\eta}G[eA^{ext}]\eta + L[A^{ext}]\right\}, \tag{3.70}$$

where $G[A^{ext}]$ denotes the fermion propagator defined in the presence of the external field,

$$G[eA^{ext}] = S_c(1 - ie\gamma \cdot A^{ext}S_c)^{-1}, \tag{3.71}$$

and

$$L = \mathrm{Tr}\,\ell n(1 - ie\gamma \cdot A^{ext}S_c)$$

$$= -i\int_0^e de' \sum_\mu \sum_{\alpha,\beta} \int d^4x \gamma_\mu^{\alpha\beta} A_\mu^{ext}(x) G_{\beta\alpha}(x,x|e'A^{ext}). \tag{3.72}$$

The very important question of the precise definition of (3.72) will be deferred until the next section. From (3.70) one sees that

$$\langle S\{A^{ext}\}\rangle = \exp L\{A^{ext}\} \tag{3.73}$$

is the probability amplitude for the vacuum to remain unchanged; hence the generating functional of this problem is simply

$$Z\{\eta,\bar{\eta}\} = \exp\left\{i\int \bar{\eta} G[eA^{ext}]\eta\right\},$$

and its functional derivatives lead to the potential theory picture of fermions propagating in the presence of A^{ext}, but having no interactions with each other. The probability of the vacuum remaining unchanged is given by $P_0 = \exp\left(2\mathrm{Re}L[A^{ext}]\right)$, which can be less than unity if the Fourier components of A_μ^{ext} can produce fermion-antifermion pairs. Because the external field can generate pairs in a variety of ways (represented by quite distinct Feynman graphs), the probability P_{2n} for producing n such pairs is somewhat more complicated than the simple Poisson form of (3.48); yet, in this model situation, it can be obtained exactly.[5]

For the completely quantized QED situation of (3.55) one has, using (3.69),

$$NZ = \exp\left[i\int \bar{\eta} S_c \left(1 - e\gamma \cdot \frac{\delta}{\delta j} S_c\right)^{-1} \eta + \mathrm{Tr}\ell n\left(1 - e\gamma \cdot \frac{\delta}{\delta j} S_c\right)\right]$$
$$\cdot \exp\left[\frac{i}{2}\int j_\mu D_c^{\mu\nu} j_\nu\right],$$

or

$$NZ = \exp\left[\frac{i}{2}\int j_\mu D_c^{\mu\nu} j_\nu\right] \cdot \exp\left[-\frac{i}{2}\int \frac{\delta}{\delta A_\mu} D_c^{\mu\nu} \frac{\delta}{\delta A_\nu}\right]$$
$$\cdot \exp\left\{i\int \bar{\eta} G[A]\eta + L[A]\right\}. \tag{3.74}$$

where use has been made of (3.61). In (3.74) the fundamental relation between potential theory and field theory is again exhibited, in the sense that one must perform specified functional differentiation operations upon the Green's functions of the former to obtain those of the latter.

E. Fields at the Same Point

An ambiguity in all the previous discussion is reflected in the lack of precision given to the definition of operator fields at the same space-time point. Such quantities are usually quite singular, while the methods of introducing a reasonable degree of finiteness are not always the same for all interactions.

The simplest level at which this problem arises concerns the definition of the free-field generating functional (3.11), obtained from the noninteracting Lagrangian (3.1), or if one uses the Schwinger construction, from (3.1) plus jA. The elementary problem now considered is the use of the action principle to obtain the generating functional for $\mu^2 \neq 0$ if it is already known for $\mu^2 = 0$. For arbitrary μ^2, one has from (3.19)

$$\frac{\delta}{\delta\mu^2} \langle 0,+\infty|0,-\infty\rangle = -\frac{i}{2}\int d^4x \, \langle 0,+\infty|A^2(x)|0,-\infty\rangle$$

$$= \frac{i}{2}\int d^4x \left(\frac{\delta}{\delta j(x)}\right)^2 \langle 0,+\infty|0,-\infty\rangle, \qquad (3.75)$$

with solution

$$\langle 0,+\infty|0,-\infty\rangle = \exp\left\{-\frac{i}{2}\int \frac{\delta}{\delta j}\,[-\mu^2]\,\frac{\delta}{\delta j}\right\} \cdot \langle 0,+\infty|0,-\infty\rangle_{\mu^2=0}, \qquad (3.76)$$

where $\langle x|\mu^2|y\rangle = \mu^2\delta(x-y)$. From (3.30),

$$\langle 0,+\infty|0,-\infty\rangle_{\mu^2=0} = \exp\frac{i}{2}\int jD_c j, \text{ where}$$

$D_c(x) = \Delta_c(x;\mu^2=0)$, while (3.56) permits the evaluation of the RHS of (3.76),

$$\exp\left[\frac{i}{2}\int jD_c(1 + [\mu^2]D_c)^{-1}j\right] \cdot \exp\left[\frac{1}{2}\,\mathrm{Tr}\ln(1 + [\mu^2]D_c)^{-1}\right]. \qquad (3.77)$$

The j-dependent term of (3.77) is exactly (3.30), since

$$<p|D_c(1 + \mu^2 D_c)^{-1}|p'> = <p|p'>\tilde{D}_c(p)\left(1 + \mu^2\tilde{D}_c(p)\right)^{-1}$$

$$= \delta^{(4)}(p-p')\left(\mu^2 + \tilde{D}_c(p)^{-1}\right)^{-1},$$

which was the starting point of the calculation. The remaining term of (3.77) is just an (infinite) phase factor,[6] but its existence rather than its value is the point requiring an explanation.

A clue to the origin of this difficulty lies in the interpretation of the phase as a vacuum-to-vacuum effect, for this change of mass shifts the energy level of the vacuum by an infinite amount. A similar effect is noted in conventional field theory, where one calculates a free-field Hamiltonian directly from the free-field Lagrangian and discards the "infinite zero-point energy" as a meaningless constant, since only energy differences can be physical. That part of the discarded constant independent of spatial momentum has returned to generate the unwanted phase of (3.77), as the self-energy of the vacuum tries to adjust to its new infinite value when μ^2 increases from zero.

The original Lagrangian may be rewritten so that it does not contain this zero-point effect by merely subtracting away the latter, replacing $L_0\{A\}$ by $L_0\{A\} - <L_0>$. Translational invariance requires that $<A^n(x)>$ be independent of x; and hence this subtraction procedure removes a constant from the term $-(\mu^2/2)A^2(x)$ of (3.1), replacing the latter by $-(\mu^2/2)[A^2(x) - <A^2(x)>]$, or for free fields, by $-(\mu^2/2):A^2(x):$. As in Chapter 2, the double dots denote a Wick-ordered product (creation operators standing on the left and destruction operators on the right). From (3.23), $A^2(x)$ is always to be understood as the symmetric limit of a time-ordered arrangement, and hence $<A^2(x)>$ may be denoted by $-i\Delta_c(0)$, a divergent constant. Application of the action principle now yields

$$\frac{\delta}{\delta\mu^2}\langle 0,+\infty|0,-\infty\rangle = -\frac{i}{2}\int dx \langle 0,+\infty|\left(A^2(x) - \langle A^2(x)\rangle\right)|0,-\infty\rangle \tag{3.78}$$

in place of (3.75), with the solution (3.77) now multiplied by an extra factor

$$\exp[\frac{i}{2}\int_0^{\mu^2} dn^2 \int dx \langle A^2(x)\rangle_{n^2}] = \exp[\frac{VT}{2}\int_0^{\mu^2} dn^2 \Delta_c(0;n^2)], \tag{3.79}$$

which exactly cancels the phase factor of (3.77). Thus, the normal ordering prescription, in providing the definition of what is meant by a pair of free field operators at the same point, restores the validity of the action principle analysis.

In a theory with nontrivial interaction of the form (3.2), one must supply a corresponding definition of n operators acting at the same point. One natural definition is obtained by the application of a Wick-ordering functional,[7]

$$W\{j\} = :\exp(i\int jA): \equiv \exp(i\int jA)_+ \langle \exp(i\int jA)_+\rangle^{-1}, \tag{3.80a}$$

where

$$:A(x_1)\cdots A(x_n): \equiv \frac{1}{i}\frac{\delta}{\delta j(x_1)}\cdots\frac{1}{i}\frac{\delta}{\delta j(x_n)} W\{j\}\Big|_{j=0}. \tag{3.80b}$$

As an example, Eqs. (3.80) applied to the interaction $L' = -(g/4!)A^4(x)$ replace the latter by

$$:L': = -\frac{g}{4!}:A^4(x): = -\frac{g}{4!}\{A^4 - \langle A^4\rangle + 6\langle A^2\rangle(A^2 - \langle A^2\rangle)\}, \tag{3.81}$$

under the assumption that $\langle A\rangle$ continues to vanish for interacting scalar fields. The use of $:L:$ rather than L is therefore equivalent to using L with the infinite constants

$$- \frac{\mu^2}{2} \langle A^2 \rangle - \frac{g}{4!} \langle A^4 \rangle - \frac{g}{4} \langle A^2 \rangle^2$$

removed, together with a shift of the unrenormalized mass μ^2 by the amount $(g/4)\langle A^2 \rangle$. Quite generally, the use of $:L:$ for any polynomial interaction of form $L' = - \sum_{i=3}^{n} g_i A^i$ corresponds to the removal of the vacuum self-energy infinity, together with the redefinition (by the addition of infinite quantities) of the unrenormalized coupling constants g_i. When $n \to \infty$, however, as for the Fradkin-Efimov[8] nonpolynomial Lagrangians, the situation is somewhat more complicated; while the normal ordering prescription may be used to remove vacuum singularities, it is not clear that the procedure maintains the proper unitarity and/or analyticity of the n-point Green's functions.[9] It should be noted that there exist other definitions of polynomial L' in which a limiting procedure for fields at the same point is employed in perturbation calculations, with the (spacelike) separation of coordinates maintained until after renormalization has been performed.[10]

For fermion fields, and in particular in QED, the situation is more complicated but more amenable to solution, since there exist certain principles which may be invoked to help define a proper interaction. Normal ordering of both $L_o\{A_\mu\}$ and $L_o\{\psi\}$ may be adopted, but this prescription is incomplete when applied to the fermion current operator of L', which has simply been written as $j_\mu = ie\bar{\psi}\gamma_\mu\psi$ in (3.49) or (3.54). Not only should the vacuum energy be zero, but in addition its current density must vanish

$$\langle j_\mu(x) \rangle_{A_\nu=0} = 0 \tag{3.82}$$

in the limit of zero external fields A_ν. Further,

for any value of A_ν, the current so induced in the vacuum by an external field must be conserved,

$$\partial_\mu \langle j_\mu(x) \rangle_{A_\nu} = 0, \tag{3.83}$$

in order to guarantee that the total charge of the vacuum remains zero. These requirements are stated for the vacuum expectation of the current operator, and consequently it is for this quantity--rather than the operator directly--that the analysis is first performed.

According to (3.23), this simplest definition of the current operator is given by

$$j_\mu(x) = i \lim_{x'-x\to 0} \sum_{\alpha,\beta} \gamma_\mu^{\alpha\beta} \left(\bar{\psi}_\beta(x')\psi_\alpha(x)\right)_+ \tag{3.84}$$

where, to maintain strict causality, the points x and x' should be relatively spacelike. In this case an alternate form of (3.84) is obtained by averaging over both orderings of these factors, to produce

$$\frac{i}{2} \lim_{x-x'\to 0} \sum_{\alpha,\beta} \gamma_\mu^{\alpha\beta} [\bar{\psi}_\beta(x'),\psi_\alpha(x)];$$

for free fermions, this is equivalent to the normal ordered form. Were these fermion fields dependent upon external, or ficticious, c-number sources A_ν, upon taking the vacuum expectation value of (3.84) there results

$$\langle j_\mu(x) \rangle_{eA} = - \lim_{x'-x\to 0} \mathrm{tr}[\gamma_\mu G(x,x'|eA)], \tag{3.85}$$

where tr means a trace over Dirac indices, and the generating functional (3.70) has been used to identify $G[A]$ with this time-ordered quantity. The relation of (3.85) to the generating functional (3.74) of QED may be seen from the relation (3.72), which with (3.85) is equivalent to

$$L[A] = i\int_0^e de'\int d^4x A_\mu(x)\langle j_\mu(x)\rangle_{e'A}. \tag{3.86}$$

An alternate and frequently useful relation may be seen to follow from these definitions,

$$\frac{\delta L}{\delta A_\mu(x)} = ie\langle j_\mu(x)\rangle_{eA}. \tag{3.87}$$

Hence any properties and requirements placed upon the current induced in the vacuum by an external field, will have an immediate application to the closed-fermion loop functional, and thereafter to all photon processes.

It should be remarked that, formally, there is no problem at all: if the fermion field equations corresponding to the interaction (3.49b) are satisfied, one expects (3.83) to be satisfied, as is (3.82) if only the limit $x - x' \to 0$ is taken in a symmetric way. The difficulty is that $G(x,x'|eA)$ becomes singular in this limit, and formal manipulations can go awry. The resolution of this problem (originally given by Valatin[11]) will here be accomplished by first changing the definition of $\langle j_\mu(x)\rangle_{eA}$ to insure that

$$\partial_\nu^y \frac{\delta}{\delta A_\nu(y)} \langle j_\mu(x)\rangle_{eA} = 0 \tag{3.88}$$

for any coordinate y, and then employing translational invariance to argue that (3.88) is indeed equivalent to (3.83).

Under gauge transformations of form $A_\mu \to A_\mu + \partial_\mu\Lambda$, the potential theory $G[A]$ will undergo the transformation

$$G(x,y|eA) \to G(x,y|e[A+\partial\Lambda])$$

$$= G(x,y|eA) \cdot \exp[ie(\Lambda(x)-\Lambda(y))], \tag{3.89}$$

as is easily seen by examining the differential equations obeyed by $G[A]$. Hence the combination

$$G_F(x,x'|A) \equiv G(x,x'|A)\exp[-ie\int_{x'}^{x} d\xi_\sigma A_\sigma(\xi)] \qquad (3.90)$$

is explicitly independent of Λ under this gauge transformation, which condition implies (Appendix J) that

$$\partial_\nu^y \frac{\delta}{\delta A_\nu(y)} G_F(x,x'|eA) = 0.$$

Thus if the $G[A]$ in the original definition of (3.85) is replaced by $G_F[A]$,

$$\langle j_\mu(x)\rangle_F \equiv - \lim_{x'-x\to 0} \mathrm{tr}[\gamma_\mu G(x,x'|eA)] \cdot \exp[-ie\int_{x'}^{x} d\xi_\sigma A_\sigma(\xi)], \qquad (3.91)$$

(3.88) will be satisfied by (3.91) for any point y and any separation $x - x'$. Taken by itself, the exponential factor of (3.91) vanishes as $x' - x \to 0$; but it multiplies $G(x,x'|eA)$ which is singular in that limit; and their product, while not itself finite, is just sufficient to insure the gauge invariance of the combination.

The nature of the limit in (3.91) should (i) be such that the coordinate difference is always spacelike, and (ii) maintains $\langle j_\mu(x)\rangle_F$ as a four-vector while (iii) preserving the translational invariance of all resulting amplitudes which arise from the further expansion of $\langle j_\mu(x)\rangle_F$ in (odd) powers of A_ν. In the perturbation expansions of QED (Chapter 6) these properties are verified *a posteriori*, while they are explicitly exhibited in the solutions of two-dimensional QED (Chapter 7). If properties (ii) and (iii) are fulfilled, then $\langle j_\mu(x)\rangle_F$ may be written in terms of its

Taylor expansion,

$$\int K^{(2)}_{\mu\nu}(x-y)A_\nu(y)dy$$

$$+ \int K^{(4)}_{\mu\nu\lambda\sigma}(x-y,y-z,z-w)A_\nu(y)A_\lambda(z)A_\sigma(w) + \cdots$$

and (3.88) employed to show that each $K^{(2\ell)}_{\mu\nu\cdots}$ satisfies $\partial^x_\mu K^{(2\ell)}_{\mu\nu\cdots} = 0$; and hence (3.83) is satisfied.

An operator redefinition of j_μ which yields (3.91) may be obtained by rewriting the interaction Lagrangian in the form

$$L' = ie\int d^4y \left(\bar{\psi}(y)\gamma_\mu\psi(x)\right)_+ A'_\mu(x,y),$$

where $A'_\mu(x,y)$ is a nonvanishing operator for spacelike values of $x-y$; e.g.,

$$A'_\mu(x,y) = A_\mu(x)f(x-y)\exp[-ie\int_y^x d\xi_\sigma A_\sigma(\xi)],$$

where $A_\mu(x)$ denotes the photon field operator or external field or their sum, and $f(x-y)$ is an arbitrary c-number function which is to approach $\delta^4(x-y)$ through spacelike values of the coordinate difference. All the analysis of the previous section goes through as before, with the $G[A']$ and $L[A']$ as defined in (3.72) and (3.74), except that now $\langle x|A'_\mu|y\rangle = A'_\mu(x,y)$. In particular, (3.72) is changed to read

$$L[A'] = -ie\int_0^1 d\lambda\int d^4x\int d^4y A'_\mu(x,y)\mathrm{tr}[\gamma_\mu G(x,y|\lambda A')], \quad (3.93)$$

while

$$G(x,y|\lambda A')$$

$$= S_c(x-y)+ie\lambda\int d^4u\int d^4v S_c(x-u)\gamma_\mu A'_\mu(u,v)G(v,y|\lambda A'). \quad (3.94)$$

Since the multipoint fermion Green's functions are not gauge invariant (in contrast to their S-matrix elements, which are), and only the photon amplitudes must satisfy these restrictions, one expects that no error will be generated by passing to the limit $f(x-y) \to \delta(x-y)$ in (3.94); in this case, the latter is equivalent to the $G[A]$ of (3.70), while (3.93) is the same as (3.86) with the replacement (3.91). Further discussion will be given within the context of the specific examples of Part II.

Notes

1. Chapter 2, Reference 2.

2. Chapter 1, Reference 3.

3. E. S. Fradkin, Nuc. Phys. 76, 588 (1966).

4. This discussion follows the treatment given by B. Zumino, NYU Lectures (1958).

5. A nonfunctional calculation of P_o, replete with combinatorics, may be found in the book by W. Thirring, *Principles of Quantum Electrodynamics*, Academic Press, Inc., London (1958). P_{2n} may be obtained by applying that formula appropriate to the fermion pair situation, analogous to (9.39), to (3.70).

6. This may be seen by supplying sufficient regularization of the propagator $\tilde{D}_c(p)$ so that the integral defining $\ell n\ N$ may be computed, after which the regularization is removed. For example, one may replace $(p^2)^{-1}$ by $2\int_0^{\Lambda^2} LdL(L+p^2)^{-3}$, and subsequently take the limit of very large Λ^2.

7. K. Symanzik, Lectures at the Summer School for High Energy Physics at Hercegnovi, Yugoslavia, August 1961.

8. The original nonpolynomial papers are due to G. V.

Efimov, JETP 44, 2107 (1963), and to E. S. Fradkin, Nuc. Phys. 49, 624 (1963). A recent review and statement of future aspirations has been given by A. Salam, "Finite Field Theory," Trieste preprint IC/71/108.

9. H. M. Fried, Nuovo Cimento 52A, 1333 (1967).

10. M. Lévy, Nuc. Phys. 57, 152 (1964); W. Zimmerman, Comm. Math. Phys. 6, 161 (1967); 10, 325 (1968).

11. J. Valatin, Proc. Royal Soc. A (London), 225, 535 and 226, 254 (1954).

Chapter 4

NONCANONICAL (CHIRAL) GENERALIZATIONS

When the interaction L' of a specific theory is dependent upon derivatives of the interacting fields, the preceding constructions must be supplemented by a careful statement and application of the ETCRs. The latter can be markedly different from the typical canonical relations discussed in Chapter 1, in which case a naive application of the formulae of Chapter 3 can be quite misleading. In this chapter, a detailed construction is given to illustrate the intricacies involved in setting up the calculations of recently relevant chiral theories.

The nonpolynomial Lagrangian[1]

$$L = -\frac{1}{2} \sum_{a,b=1}^{3} \sum_{\mu} \partial_\mu \pi_a \cdot F_{ab}(\pi) \cdot \partial_\mu \pi_b \tag{4.1}$$

is written to describe zero-mass pions, of isotopic index $\underline{a}$; here $F_{ab} = \delta_{ab} f_1(\vec{\pi}^2) + \pi_a \pi_b f_2(\vec{\pi}^2)$, with the functions $f_{1,2}$ given in terms of

$$f_1(\vec{\pi}^2) = f_\pi^2 (f^2 + \vec{\pi}^2)^{-1}, \tag{4.2a}$$

$$f_2(\vec{\pi}^2) = f_\pi^2 (f^2 + \vec{\pi}^2)^{-2} (4\vec{\pi}^2 f'^2 - 4ff' - 1), \tag{4.2b}$$

where $f(\vec{\pi}^2)$ denotes the so-called "pion gauge function," $f' \equiv \frac{d}{d\pi^2} f(\pi^2)$, and $f_\pi^{-1} = g$ plays the role of a coupling constant. In a theory possessing strict chiral invariance any functional form is allowed for f, as long as the hermiticity of (4.1) is maintained. If $F_{ab} = \delta_{ab} + \overline{F}_{ab}$, one might simply write (4.1) as a "free-field" part (F_{ab} replaced by δ_{ab}) plus an "interaction" part (F_{ab} replaced by $\overline{F}_{ab}$); and one might then apply the previous techniques in which the complete generating functional is given in terms of "interaction" functional differentiation operations upon the free-field generating functional. What is wrong with such a program is that proper account is not taken of the correct definition of generalized momenta,

$$P_a \equiv \frac{\partial L}{\partial(\partial_o \pi_a)} = F_{ab}(\pi)\partial_o \pi_b \tag{4.3}$$

and the corresponding ETCR,

$$[\pi_a(x), P_b(y)]_{x_o=y_o} = i\delta_{ab}\delta^3(\overrightarrow{x-y}), \tag{4.4}$$

or

$$[\pi_a(x), \partial_o \pi_b(y)]_{x_o=y_o} = i\delta^3(\overrightarrow{x-y})F^{-1}_{ab}[\pi(x)], \tag{4.5}$$

where (4.5) follows from (4.4) and the unchanged relation (1.2), $[\pi_a(x), \pi_b(y)]_{x_o=y_o} = 0$.

The quantization of this Lagrangian is of interest in the understanding of theories which are expected to satisfy certain (in this case, chiral) symmetry requirements, such as the vanishing of the pion's mass in the presence of self-interactions written (as in (4.1)) to preserve the symmetry. On the basis of the naive, or improper, calculation, the retention of such symmetry properties is not at all clear; in fact, it was noticed that the first few leading, highly divergent perturbation contributions to the pion self-mass could cancel, in each order, if a special "gauge" or choice of $f(\pi^2)$ was made. This is contrary to the spirit of chiral invariance for arbitrary f; in terms of F_{ab}, the condition for such cancellation was found to be $\det F = 1$. However, when a proper computation is performed, taking into account the change in the generating functional that exists in such a noncanonical theory, one finds that the cancellation occurs automatically, for any f, and in this sense preserves the chiral property of zero pion mass.

The calculation given here represents a slight generalization of the Symanzik construction, but an entirely analogous construction in the Schwinger manner may be carried through. Variation of (4.1) yields the field equations

$$\partial_\mu[F_{ab} \cdot \partial_\mu \pi_b] = \frac{1}{2}\, \partial_\mu \pi_c \cdot \frac{\partial F_{cd}}{\partial \pi_a} \cdot \partial_\mu \pi_d,$$

or

$$\partial_o[F_{ab} \cdot \partial_o \pi_b] = \partial_j[F_{ab} \cdot \partial_j \pi_b] - \frac{1}{2}\, \partial_j \pi_c \cdot \frac{\partial F_{cd}}{\partial \pi_a} \cdot \partial_j \pi_d$$

$$+ \frac{1}{2}\, \partial_o \pi_c \cdot \frac{\partial F_{cd}}{\partial \pi_a} \cdot \partial_o \pi_d. \tag{4.6}$$

In terms of the generalized momenta P_a of (4.3), the last term on the RHS of (4.6) may be replaced by

$$\frac{1}{2}\, P_\varepsilon F^{-1}_{\varepsilon r} \cdot \frac{\partial F_{rs}}{\partial \pi_a} \cdot F^{-1}_{st} P_t = -\frac{1}{2}\, P_\varepsilon \cdot \frac{\partial F^{-1}_{\varepsilon r}}{\partial \pi_a} \cdot P_r, \tag{4.7}$$

while the LHS is simply $\partial_o P_a$.

Because the interaction part of the equations of motion involve P_a, it is clear that the generating functional must also. One defines

$$Z\{j,k\} = \langle (\exp i\!\int [j_a \pi_a + k_a P_a])_+ \rangle, \tag{4.8}$$

where $\vec{j}(x)$ and $\vec{k}(x)$ are appropriate c-number, isotopic sources; at the end of the calculation, $\vec{k}$ will be set equal to zero, restoring the manifest Lorentz covariance of $Z\{j,0\}$. In exactly the same manner as described in the Symanzik construction, one obtains the pair of equations

$$\partial_o \frac{1}{i}\frac{\delta}{\delta j_a(x)} Z = - k_a(x)Z + F^{-1}_{ab}\left(\frac{1}{i}\frac{\delta}{\delta j(x)}\right) \cdot \frac{1}{i}\frac{\delta}{\delta k_b(x)} Z, \tag{4.9}$$

and

$$\partial_o \frac{1}{i}\frac{\delta}{\delta k_a(x)} Z = + j_a(x)Z + \{\partial_j [F_{ab}\left(\frac{1}{i}\frac{\delta}{\delta j(x)}\right)\left(\partial_j \frac{1}{i}\frac{\delta}{\delta j_b(x)}\right)]$$

$$- \frac{1}{2}\left(\partial_j \frac{1}{i}\frac{\delta}{\delta j_c(x)}\right)\frac{\partial F_{cd}}{\partial \pi_a}\left(\frac{1}{i}\frac{\delta}{\delta j(x)}\right)\left(\partial_j \frac{1}{i}\frac{\delta}{\delta j_d(x)}\right)$$

$$- \frac{1}{2}\left(\frac{1}{i}\frac{\delta}{\delta k_c(x)}\right)\frac{\partial F^{-1}_{cd}}{\partial \pi_a}\left(\frac{1}{i}\frac{\delta}{\delta j(x)}\right)\left(\frac{1}{i}\frac{\delta}{\delta k_d(x)}\right)\}Z. \tag{4.10}$$

The first term on the RHS of both (4.9) and (4.10) is "kinematical," coming from the ETCR (4.4), while the remaining content of these relations is similar to those of the Hamiltonian equations, expressing the time dependence of the fields and generalized momenta, respectively.

Rather than adopting the previous Symanzik-type ansatz, it will be useful to exhibit an alternate method of construction of Z, one which proceeds from the concept of a functional Fourier transform. Any finite polynomial $F_n\{j\}$ can always be written in terms of suitable functional derivatives with respect to $\chi(x)$ of the quantity $\exp(i\int j\chi)$; e.g.,

$$\int j(x_1)\cdots j(x_n)\Xi(x_1,\cdots x_n)$$

$$= \int \Xi(x_1,\cdots x_n)\, \frac{1}{i}\frac{\delta}{\delta\chi(x_1)} \cdots \frac{1}{i}\frac{\delta}{\delta\chi(x_n)} (\exp\, i\int j\chi)\Big|_{\chi=0}. \tag{4.11}$$

A compact way of stating this is to invent notation and define manipulation analogous to those used in conjunction with ordinary δ-functions; that is, (4.11) may be replaced by

$$\int d[\chi]\tilde{F}_n\{\chi\}\exp(i\int j\chi) \tag{4.12}$$

with

$$\tilde{F}_n\{\chi\} \equiv \int \Xi(x_1,\cdots x_n)\left(-\frac{1}{i}\frac{\delta}{\delta\chi(x_1)}\right)\cdots\left(-\frac{1}{i}\frac{\delta}{\delta\chi(x_n)}\right)\delta[\chi],$$

and the rule

$$\int d[\chi]G\{\chi\}\frac{\delta}{\delta\chi(x)}\delta[\chi] = -\left.\frac{\delta G}{\delta\chi(x)}\right|_{\chi=0}.$$

For certain $F\{j\}$, it may be possible to treat the sum of such an infinite sequence in the same way as every term of the sum, and so express a nonpolynomial $F\{j\}$ in terms of a functional Fourier transform,

$$F\{j\} = \int d[\chi]\tilde{F}\{\chi\}\exp(i\int j\chi). \tag{4.13}$$

This can be done if $F\{j\}$ is of Gaussian form, which is all that shall be required in this discussion; in fact, with the aid of (3.56), (4.13) may be readily evaluated if $\tilde{F}\{\chi\}$ is Gaussian, without ever calculating an integral. In Appendix K it is shown that

$$\begin{aligned} Q_\pm[j;B] &\equiv \int d[\chi]\exp[i\int j_a\chi_a \mp \frac{i}{2}\int \chi_a B_{ab}\chi_b] \\ &= C_\pm \exp[\pm\frac{i}{2}\int j_a \cdot B^{-1}_{ab} \cdot j_b + \frac{1}{2}\mathrm{Tr}\,\ln B^{-1}] \end{aligned} \tag{4.14}$$

where the constant $C_\pm \equiv Q_\pm[0;1]$ is independent of j and B, and Tr includes a sum over isotopic indices.

Following this idea, a transform in two variables is introduced for $Z\{j,k\}$,

$$Z\{j,k\} = \int d[\chi]\int d[\psi]\tilde{Z}[\chi,\psi]\exp i\int[j_a\cdot\chi_a+k_a\cdot\psi_a], \qquad (4.15)$$

in order that the differential equations (4.9), (4.10) may be converted into a simpler set of relations satisfied by $\tilde{Z}[\chi,\psi]$,

$$\partial_o\chi_a(x)\cdot\tilde{Z} = -i\frac{\delta}{\delta\psi_a(x)}\tilde{Z} + F^{-1}_{ab}(\chi(x))\psi_b(x)\cdot\tilde{Z}, \qquad (4.16)$$

and

$$\partial_o\psi_a(x)\cdot\tilde{Z} = i\frac{\delta}{\delta\chi_a(x)}\tilde{Z} + \{\partial_j[F_{ab}(\chi)\partial_j\chi_b] - \frac{1}{2}(\partial_j\chi_c)\frac{\partial F_{cd}}{\partial\chi_a}(\chi)(\partial_j\chi_d) - \frac{1}{2}\psi_c\cdot\frac{\partial F^{-1}_{cd}}{\partial\chi_a}(\chi)\cdot\psi_d\}\tilde{Z}. \qquad (4.17)$$

A simple quadrature of (4.16) yields

$$\tilde{Z}[\chi,\psi] = N[\chi]\exp[i\int\psi_a\cdot\partial_o\chi_a - \frac{i}{2}\int\psi_a\cdot F^{-1}_{ab}(\chi)\cdot\psi_b], \qquad (4.18)$$

where $N[\chi]$ remains to be determined. After an integration-by-parts on the first exponential factor of (4.18), substitution into (4.17) provides the necessary, and simple, equation for $N[\chi]$,

$$i\frac{\delta}{\delta\chi_a(x)}\ell n\, N[\chi] = \frac{1}{2}(\partial_j\chi_c)\frac{\partial F_{cd}}{\partial\chi_a}(\chi)(\partial_j\chi_d) - \partial_j[F_{ab}(\chi)\partial_j\chi_b], \qquad (4.19)$$

with solution

$$N[\chi] = N_o \exp[-\frac{i}{2}\int(\partial_j\chi_c)F_{cd}(\chi)(\partial_j\chi_d)], \qquad (4.20)$$

where N_o is a constant. In all terms, $F_{cd}[\chi(x)]$ is understood as dependent upon the single configuration coordinate, x. If the formal, multiple integral notation is used, then $\langle x|F_{ab}|y\rangle = \delta(x-y)F_{ab}[\chi(x)]$.

Equations (4.20) and (4.18) may now be inserted into the representation (4.15) for $Z\{j,k\}$, and the functional integral over ψ_a calculated with the aid of (4.14),

$$Z\{j\} = C_+N_o\int d[\chi]\exp\,[i\int j_a\cdot\chi_a - \frac{i}{2}\int(\partial_\mu\chi_a)\cdot F_{ab}(\chi)\cdot$$

$$\cdot\,(\partial_\mu\chi_b) + \frac{1}{2}\,\mathrm{Tr}\,\ln F(\chi)], \qquad (4.21)$$

in which k_a, no longer necessary, has been set equal to zero. Equation (4.21) may be put into the form of Chapter 3 by splitting F into a free- and interaction-part, $F_{ab} = \delta_{ab} + \overline{F}_{ab}$, and rewriting all the interaction dependence in terms of operations upon a gaussian integral,

$$Z\{j\} = C_+N_o\exp[\frac{1}{2}\,\mathrm{Tr}\,\ln F\left(\frac{1}{i}\frac{\delta}{\delta j}\right) - \frac{i}{2}\int\left(\partial_\mu\frac{1}{i}\frac{\delta}{\delta j_a}\right)\cdot$$

$$\cdot\,\overline{F}_{ab}\left(\frac{1}{i}\frac{\delta}{\delta j}\right)\left(\partial_\mu\frac{1}{i}\frac{\delta}{\delta j_b}\right)]\,\cdot$$

$$\cdot\int d[\chi]\exp[i\int j_a\chi_a - \frac{i}{2}\int\partial_\mu\chi_a\cdot\partial_\mu\chi_a]. \qquad (4.22)$$

Another integration-by-parts converts the quadratic exponential dependence under this functional integral to $-\frac{i}{2}\int\chi_a(-\partial^2)\chi_a$; and when now applying (4.14), the operator $(-\partial^2)^{-1}$ is understood to be

$$(-\partial^2 - i\varepsilon)^{-1}\Big|_{\varepsilon\to 0+} = D_c,$$

since the construction is of a time-ordered quantity. Hence, (4.22) may be written as

$$NZ\{j\} = \exp[\tfrac{1}{2}\,\mathrm{Tr}\,\ln F\left(\frac{1}{i}\frac{\delta}{\delta j}\right) - \frac{i}{2}\int\left(\partial_\mu \frac{1}{i}\frac{\delta}{\delta j_a}\right)\overline{F}_{ab}\left(\frac{1}{i}\frac{\delta}{\delta j}\right)\cdot$$

$$\cdot\left(\partial_\mu \frac{1}{i}\frac{\delta}{\delta j_b}\right)]\cdot\exp[\tfrac{i}{2}\int j_a\cdot D_c\cdot j_a], \qquad (4.23)$$

where the overall normalization constant has simply been written as $N^{-1} = C_+^2 N_o \exp[\frac{1}{2}\,\mathrm{Tr}\,\ln D_c]$. The alternate, and somewhat more convenient form following from (3.61) is

$$NZ\{j\} = \exp[\tfrac{i}{2}\int j_a\cdot D_c\cdot j_a]\cdot\exp[-\frac{i}{2}\int\frac{\delta}{\delta\pi_a}D_c\frac{\delta}{\delta\pi_a}]\cdot$$

$$\cdot\exp[\tfrac{1}{2}\,\mathrm{Tr}\,\ln F(\pi) - \frac{i}{2}\int(\partial_\mu\pi_a)\overline{F}_{ab}(\pi)(\partial_\mu\pi_b)], \qquad (4.24)$$

where

$$\pi_a(x) \equiv \int d^4y\, D_c(x-y)j_a(y),$$

Had the canonical rules for the construction of $Z\{j\}$ been (mis)applied to this problem, the result would have been (4.23) or (4.24) without the factor $\exp[\frac{1}{2}\,\mathrm{Tr}\,\ln F]$, which term would then be missing in any perturbation calculation. By Appendix G, this term is equivalent to $\exp[\frac{1}{2}\,\delta^4(0)\int d^4x\,\ln\det F(x)]$. If the chiral invariance of the theory is to be maintained for $\overline{F}\not\equiv 0$, the highly divergent perturbation terms independent of external momenta, which are generated by $\exp[\frac{1}{2} \text{Tr } \ell n F]$, must be canceled by self-processes arising from the remainder of the interaction. Hence, if one neglects the $\exp[\frac{1}{2} \text{Tr } \ell n F]$ dependence, the only way to remove the remaining (and leading) divergences will be by a special choice of pion gauge, equivalent to the requirement det $F = 1$. In fact, the complete generating functional of (4.24) does maintain chiral invariance for any f, as illustrated by recent lowest-order calculations of the pion self-energy.[2]

Notes

1. This problem was discussed by J. M. Charap, Phys. Rev. D2, 1554 (1970), where references to appropriate chiral literature may be found. In the context of general relativity, results equivalent to (4.23) were first given by B. S. DeWitt, Physical Review Letters 12, 742 (1964). A very lucid summary of functional applications to related topics, including gravitation and Yang-Mills theories, has been given by E. Fradkin, U. Esposito and S. Termini, Rivista del Nuovo Cimento, Serie I, Vol. 2, 498 (1970).

2. I. Gerstein, R. Jackiw, B. Lee, and S. Weinberg, Phys. Rev. D3, 2486 (1971); and J. M. Charap, Phys. Rev. D3, 1998 (1971).

Chapter 5

SPECIAL TOPICS IN QUANTUM ELECTRODYNAMICS

A. The Heavy Proton Limit

In previous discussions, an early distinction has been made between external, c-number fields, and the quantized fields which are of paramount interest here. One has the intuitive understanding that classical fields are appropriate for those limiting situations where the exact fluctuations of a quantized field are small and unimportant; and indeed, such idealizations are contained in the conceptual foundations of measurement theory. In this section, such a limit will be exhibited, as the completely quantized theory of protons, electrons and photons is shown to reduce, in the limit of very large proton mass, to a completely quantized theory of photons and electrons moving in the presence of external Coulomb fields generated by charges located at the positions of the massive protons.[1]

The presentation employed here will first establish the essential part of this limit in a somewhat simplified situation, involving but one electron and one proton interacting by the exchange of virtual photons, and will then be extended to the general case, without approximation. The configuration space amplitude for the exchange of all possible photons between an electron and proton is given by the functional statement

$$M(x,y;x',y') = \exp\left[-i\int \frac{\delta}{\delta A_\mu} D_c^{\mu\nu} \frac{\delta}{\delta A'_\nu}\right] \cdot G_p(x,y|A) \cdot$$

$$\cdot\, G_e(x',y'|A')\Big|_{A=A'=0}, \tag{5.1}$$

written in exact analogy with (3.66). Since the indicated functional operation is a translation of either field, an equivalent form is

$$M = G_p\left[x,y\Big|-i\int D_c \frac{\delta}{\delta A'}\right] \cdot G_e(x',y'|A')\Big|_{A'=0}, \tag{5.2}$$

where $G_p(x,y|A)$ denotes the relativistic, potential

theory Green's function of the proton in the presence of a ficticious, external, field A; and similarly for the electron's Green's function $G_e[A']$. For simplicity, the discussion is carried out in the Feynman gauge, $D_c^{\mu\nu} = \delta_{\mu\nu} \cdot D_c$.

The essence of the procedure follows from the observation that in the limit of very large proton mass M, the proton Green's function takes on a form sufficiently simple to be able to perform the operations of (5.2). It is

$$G_p(x,y|A) = i\theta(x_o - y_o)\delta(\overrightarrow{x-y}) \cdot$$
$$\cdot \exp[-iM(x_o - y_o) - ie\int_{y_o}^{x_o} d\xi A_o(\vec{x},\xi)], \qquad (5.3)$$

where $A_o = -iA_4$. Derivation and application of forms similar to (5.3) will be discussed fully in Chapters 8 and 9, but for the present purpose one can see that (5.3) represents a solution to the approximate equation

$$(M + [\frac{1}{i}\partial_o^x - ieA_4(x)])G_p(x,y|A) = \delta^4(x-y), \qquad (5.4)$$

and its adjoint. It has been obtained from the exact differential equation satisfied by (3.71) by the neglect of all $\vec{\gamma}$-dependence, and the replacement $\gamma_4 = +1$; these approximations represent the zero velocity appropriate to an infinitely massive proton, and to the "propagation" of a massive proton rather than a massive anti-proton. (Actually, a more proper statement is that the heavy proton's velocity should be constant, rather than necessarily zero, a refinement treated in Chapter 8.)

With (5.3), the functional operations of (5.2) are simply those of another translation, and yield

$$M(x,y;x',y') = i\theta(x_o - y_o)\delta(\overrightarrow{x-y})\exp[-iM(x_o - y_o)] \cdot$$

$$\cdot \; G_e(x',y'|A^{eff}), \tag{5.5}$$

where

$$A_\mu^{eff}(z) = ie\delta_{\mu 4} \int_{y_o}^{x_o} d\xi D_c(\overrightarrow{x-z}, \xi - z_o) \tag{5.6}$$

defines the effective, position dependent field seen by the electron. Now, in the limit $x_o \to +\infty$, $y_o \to -\infty$, which physically corresponds to the continued presence of the heavy proton, the integral of (5.6) may be easily evaluated to give

$$A_\mu^{eff}(z) = i\delta_{\mu 4}\left(\frac{e}{4\pi}\right)|\overrightarrow{z-x}|^{-1},$$

just the Coulomb field acting on the electron at $\vec{z}$ due to the proton at $\vec{x}$. The remaining factors of (5.3), which simply multiply $G_e(x',y'|A^{eff})$ represent the wave function of a fixed, positive energy proton, with the energy dependence here contributing a phase subsequently removed upon calculating the probability for any process involving the electron.

The generalization of this observation, without approximation, is not difficult. From the complete interaction for quantized protons, electrons and photons,

$$\mathcal{L}' = iA_\mu[e\,\bar{\psi}_p\gamma_\mu\psi_p + e\,\bar{\psi}_e\gamma_\mu\psi_e]$$

(where the limiting behavior of the previous chapter which defines the proton and electron currents is suppressed), one constructs the complete generating functional

$$N(e+p)\mathcal{Z}\{j_\mu,\bar{\eta}_e,\eta_e,\bar{\eta}_p,\eta_p\}$$

$$= \exp[i\int\bar{\eta}_e G_e\left(\frac{1}{i}\frac{\delta}{\delta j}\right)\eta_e + i\int\bar{\eta}_p G_p\left(\frac{1}{i}\frac{\delta}{\delta j}\right)\eta_p + L_e\left(\frac{1}{i}\frac{\delta}{\delta j}\right) +$$

$$+ L_p\left(\frac{1}{i}\frac{\delta}{\delta j}\right)] \cdot \exp[\frac{i}{2}\int jD_c j].$$

Functional differentiation with respect to η_p and $\bar{\eta}_p$, followed by the replacements $\eta_p = \bar{\eta}_p = 0$, yields an exact statement for the remnant of the generating functional representing a single proton and arbitrary numbers of electrons and photons,

$$N(e+p)Z_{1p} = \exp[\frac{i}{2}\int jD_c j]\cdot\exp[-\frac{i}{2}\int \frac{\delta}{\delta A} D_c \frac{\delta}{\delta A}]\cdot G_p(x,y|A)\cdot$$

$$\cdot \exp\left(i\int\bar{\eta}_e G_e[A]\eta_e + L_e[A] + L_p[A]\right), \qquad (5.7)$$

where use has been made of (3.61). With the aid of (3.67), this can be put into the convenient form

$$N(e+p)Z_{1p} = \exp[\frac{i}{2}\int jD_c j]\cdot\exp[-i\int \frac{\delta}{\delta A_1} D_c \frac{\delta}{\delta A_2}]\cdot$$

$$\cdot\left\{\exp[-\frac{i}{2}\int \frac{\delta}{\delta A_1} D_c \frac{\delta}{\delta A_1}]\cdot G_p(x,y|A_1)\cdot\right.$$

$$\left.\cdot\exp L_p[A_1]\right\}\cdot\left\{\exp\left(-\frac{i}{2}\int \frac{\delta}{\delta A_2} D_c \frac{\delta}{\delta A_2}\right)\cdot\right.$$

$$\left.\cdot\exp\left(i\int\bar{\eta}_e G_e[A_2]\eta_e + L_e[A]\right)\right\}\Big|_{A_1=A_2=A}. \qquad (5.8)$$

Because the approximate proton propagator of (5.3) is a retarded function in configuration space, it follows that $L_p[A]$, defined in terms of trace operations, must vanish; and in this case, $N(e+p) \to N(e)$. Further, the effect of the self-linkage operator on the proton Green's function will be to multiply the latter by a factor

$$\exp[-\frac{i}{2}e^2\int_{y_o}^{x_o} d\xi \int_{y_o}^{x_o} d\eta\, D_c(\vec{0};\xi-\eta)]. \qquad (5.9)$$

Using the symmetry of $D_c(x-y)$ under interchange of its arguments, (5.9) may be rewritten as

$$- \frac{i}{2} e^2(x_o-y_o)(2\pi)^{-3} \int \frac{d^3k}{\vec{k}^2}$$

$$- \frac{ie^2}{(2\pi)^4} \int \frac{d^4k}{k^2+\mu^2-i\varepsilon}\left(\frac{1}{k_o-i\varepsilon}\right)^2[1- e^{-ik_o(x_o-y_o)}], \quad (5.10)$$

where dependence on a small "photon mass" has been introduced in order to avoid the infrared difficulties which almost always require special attention; the two ways in which that problem is circumvented--retaining a small photon mass until the last step of any S-matrix calculation, or the use of the Yennie gauge in which the asymptotic condition for Fermion fields is not explicitly violated--will be discussed in Chapter 9. Here, it is sufficient to note that (5.10) takes the simple form $-i\delta M(x_o-y_o) + \ell n\, Z_2^{(p)}$ in the limit $x_o-y_o \to \infty$, where δM and $Z_2^{(p)}$ are divergent constants, easily recognized in (5.10), to which an interpretation must be given: $Z_2^{(p)}$ renormalizes (i.e. multiplies) the proton's wave function by a (gauge-dependent) factor, thereby expressing the existence of all self-processes; while δM is a mass shift of the heavy proton due to a self-interaction with its photon cloud. The combination $M+\delta M = m_p$ will be treated as the physical, or renormalized proton mass, and (5.8) written as

$$Z_{1p}\Big|_{x_o-y_o\to\infty} \to iZ_2^{(p)}\delta(\overrightarrow{x-y})e^{-im_p(x_o-y_o)} Z\{j_\mu,\bar{\eta}_e,\eta_e;A^{eff}\}, \quad (5.11)$$

where $Z\{j_\mu,\bar{\eta}_e,\eta_e;A^{eff}\}$ denotes the exact generating functional for interacting electrons and photons moving in the Coulomb field (5.6),

$$N(e)Z\{j,\bar{\eta}_e,\eta_e;A^{eff}\} = \exp[\frac{i}{2}\int jD_c j]\cdot\exp[-\frac{i}{2}\int \frac{\delta}{\delta A} D_c \frac{\delta}{\delta A}]\cdot$$

$$\cdot\ \exp\{i\int\bar{\eta}G[A+A^{eff}]\eta+L_e[A+A^{eff}]\}. \tag{5.12}$$

The generalization of (5.12) to several massive protons is immediate. A far more difficult task is the development of proton recoil corrections to this limiting case, especially when the effects of strong interactions are included.[2]

B. Green's Function Equations

In the first sentence of Chapter 1 it was stated that quantum field theory may be described as a theory of coupled Green's functions and it is most appropriate to exhibit such relations directly from the formal solutions for the generating functional. The latter have been constructed using the input information of field equations plus equal-time-commutation-relations, and are therefore equivalent to, if somewhat more convenient than the more conventional derivations of the relations between multi-point Green's functions.

One begins by defining the different n-point functions

$$S_c'(x-y) = i\langle(\psi(x)\bar{\psi}(y))_+\rangle, \tag{5.13a}$$

$$V_\mu(x,y;z) = \langle(\psi(x)\bar{\psi}(y)A_\mu(z))_+\rangle, \tag{5.14a}$$

$$M_{\mu\nu}(x,y;z_1,z_2) = \langle(\psi(x)\bar{\psi}(y)A_\mu(z_1)A_\nu(z_2))_+\rangle, \tag{5.15a}$$

etc., and giving their corresponding functional representations,

$$S_c'(x-y) = G(x,y|\frac{1}{i}\frac{\delta}{\delta j})N^{-1}\ \exp\ L[\frac{1}{i}\frac{\delta}{\delta j}]\ \cdot$$

$$\cdot \exp\left[\frac{i}{2}\int jD_c j\right]\Big|_{j=0}, \tag{5.13b}$$

$$V_\mu(x,y;z) = \frac{1}{i}\frac{\delta}{\delta j_\mu(z)} G\left(x,y\Big|\frac{1}{i}\frac{\delta}{\delta j}\right)N^{-1} \exp L\left[\frac{1}{i}\frac{\delta}{\delta j}\right]\cdot$$

$$\cdot \exp\left[\frac{i}{2}\int jD_c j\right]\Big|_{j=0}, \tag{5.14b}$$

$$M_{\mu\nu}(x,y;z_1,z_2) = \frac{1}{i}\frac{\delta}{\delta j_\mu(z_1)}\cdot\frac{1}{i}\frac{\delta}{\delta j_\nu(z_2)} G\left(x,y\Big|\frac{1}{i}\frac{\delta}{\delta j}\right)N^{-1}\cdot$$

$$\cdot \exp L\left[\frac{1}{i}\frac{\delta}{\delta j}\right]\cdot \exp\left[\frac{i}{2}\int jD_c j\right]\Big|_{j=0}. \tag{5.15b}$$

The prime on S_c' indicates a completely "dressed," although unrenormalized fermion propagator, which by translational invariance is a function of the difference of its configuration space coordinates; the same remark applies to V_μ, $M_{\mu\nu}$ and all the other n-point Green's functions. With the aid of the integral equation, itself just an expression of (3.71),

$$G\left(x,y\Big|\frac{1}{i}\frac{\delta}{\delta j}\right) = S_c(x-y) + ie\int S_c(x-z)\gamma\cdot$$

$$\cdot\frac{1}{i}\frac{\delta}{\delta j(z)} G\left(z,y\Big|\frac{1}{i}\frac{\delta}{\delta j}\right)d^4z, \tag{5.16}$$

relations between each of these Green's functions may be obtained. For example, (5.13b) can be converted into

$$S_c'(x-y) = S_c(x-y) + ie\int S_c(x-z)\gamma_\mu V_\mu(z,y;z), \tag{5.13c}$$

when use is made of (5.14b) and the definition of N. In a similar way, taking into account the dependence of $L[A]$ on even powers of A, one finds

$$V_\mu(x,y;z) = ie \int S_c(x-z')\gamma_\sigma M_{\sigma\mu}(z',y;z,z') \qquad (5.14c)$$

while an analogous relation exists expressing the four-point function $M_{\mu\nu}$ in terms of S_c, the dressed photon propagator $D'_{c,\mu\nu}$, and the five-point function of two external fermion and three boson lines $M_{\mu\nu\lambda}$. The resolution of these equations is the central problem of relativistic field theory; and all of the models presently known may be understood as approximations to this sequence of exact Green's function equations.

There are two aspects concerning the manner in which n-point functions acquire structure which are worth mentioning in this discussion, the generic form of the two-point propagator functions, and the way in which the latter enter into the structure of all the higher n-point functions. The propagators in QED, S'_c and $D'_{c,\mu\nu}$ may be written in terms of the very-well-known Lehman-Källén representation; the latter is most easily derived for a scalar, Boson field[3]

$$\Delta'_c(x-y) = i\langle (A(x)A(y))_+\rangle = \int_0^\infty d\kappa^2 \rho(\kappa^2)\Delta_c(x-y;\kappa^2) \qquad (5.17a)$$

or its momentum space transform,

$$\tilde{\Delta}'_c(p) = \int_0^\infty d\kappa^2 \rho(\kappa^2)(\kappa^2+p^2-i\varepsilon)^{-1}, \qquad (5.17b)$$

where the spectral function $\rho(\kappa^2)$ is defined by

$$\rho(-P_n^2) = |{}_{IN}\langle 0|A(0)|n\rangle_{IN}|^2 \geq 0, \qquad (5.18)$$

and $\Delta_c(x;\kappa^2)$ denotes the free particle propagator of mass κ^2. In (5.18), P_n denotes the total four-momentum of the n-particle state $|n\rangle_{IN}$. Aside from questions concerning the existence of the integrals in

(5.17), this representation follows from translational and Lorentz invariance. In a theory containing a single mass m, identified with the renormalized, or physical mass of the particle one is trying to describe, $\rho(\kappa^2)$ takes the form

$$\rho(\kappa^2) = Z\delta(\kappa^2-m^2) + \theta(\kappa - [m+\delta])\sigma(\kappa^2), \tag{5.19}$$

where Z denotes the wave function renormalization constant, and $\sigma(\kappa^2)$ is the spectral function of relevance for κ^2 values exceeding the continuum threshold $(m+\delta)^2$. A relation similar to (5.17) holds for every two-point function; in particular, the dressed commutator may be represented in terms of a similar integral over the free-particle commutator of mass κ^2,

$$\Delta'(x-y) = \int_0^\infty d\kappa^2\ \rho(\kappa^2)\Delta(x-y;\kappa^2). \tag{5.20}$$

Application of the ETCR (1.3) then generates the sum rule

$$1 = \int_0^\infty d\kappa^2\ \rho(\kappa^2), \tag{5.21}$$

showing that $0 \leq Z \leq 1$. In model calculations, one typically finds divergences in the computations for Z^{-1}, but these constants have the pleasant property of being removed in the redefinitions which accompany renormalization. Similar representations hold for the transverse part of the photon propagator (defined in the next section) and for the fermion propagator (here a pair of spectral functions is required), with constants Z_3 and Z_2, respectively; the reader is referred to the very standard and numerous discussions which exist on this interesting subject.[3] The observation to be noted here concerns the limit

$$\lim_{x-y\to\infty} \Delta_c'(x-y) \sim Z \lim_{x-y\to\infty} \Delta_c(x-y;m^2) + \cdots \tag{5.22}$$

where the neglected terms on the RHS of (5.22) vanish faster than the exhibited free-particle propagator, if δ of (5.19) is > 0. (An example of this may be seen in (5.10).)

The second observation of interest here is expressed by the statement that every connected n-point Green's function, $n \geq 3$, has a dressed propagator of appropriate particle type at every leg, or external line; e.g., the momentum space vertex function of QED may be written as

$$V_\mu(p,p';k) = \tilde{S}_c'(p)\Gamma_\nu(p,p';k)\tilde{S}_c'(p') \cdot \tilde{D}'_{c,\mu\nu}(k),$$

with Γ_ν the sum over all proper vertex parts (those not containing the self-energy structure already in each $\tilde{S}_c'$ and $\tilde{D}_c'$). Each such propagator contains a one-particle-of-physical-mass pole term, multiplied by its wave function renormalization constant, plus the continuum part of the propagator; and only the pole contribution of each propagator remains upon calculating the mass-shell, amputated, S-matrix elements, as in (2.29).

C. Gauge Transformations and the Ward Identity

The generating functional of (3.74) has been discussed in the Feynman Gauge, where $D_{c,\mu\nu} = \delta_{\mu\nu}D_c$. In this section, the so-called general relativistic gauge in which

$$D_{c,\mu\nu}(x) \to D_{c,\mu\nu}(x) + \partial_\mu\partial_\nu M_c(x),$$

with $M_c(x)$ an arbitrary, causal, invariant function of x^2, will be studied in order to be able to define[4] "gauge transformations of the third kind," the changes in the multipoint Green's functions, as M_c changes to $M_c + \Delta M_c$, with ΔM_c not necessarily small. The physical content of the theory remains unchanged by such transformations, although certain calculations are simplified in a specific gauge.

It is convenient to begin with the free photon propagator written in the general form

$$D_{c,\mu\nu} = \Pi_{\mu\nu} D_c + \partial_\mu \partial_\nu M_c$$

where $\Pi_{\mu\nu}$ denotes the transverse projection operator

$$\delta_{\mu\nu} - \partial_\mu \partial_\nu (\partial^2 + i\varepsilon)^{-1}; \quad D^{(T)}_{c,\mu\nu} = \Pi_{\mu\nu} D_c \quad \text{and} \quad D^{(L)}_{c,\mu\nu} = \partial_\mu \partial_\nu M_c$$

are then the transverse and longitudinal parts of the propagator, respectively. This decomposition is quite general, and may always be effected with the aid of the projection operator which extracts the transverse part of any symmetric tensor $Q_{\mu\nu}$ upon which it operates, $Q^{(T)}_{\mu\nu} = \sum_\lambda \Pi_{\mu\lambda} Q_{\lambda\nu}$. In such an arbitrary gauge, the generating functional of QED may be written as

$$\begin{aligned} NZ^{(M)} &= \exp\left\{ i\int \bar{\eta} G[\tfrac{1}{i}\tfrac{\delta}{\delta j}]\eta + L[\tfrac{1}{i}\tfrac{\delta}{\delta j}] \right\} \cdot \\ &\quad \cdot \exp[-\tfrac{i}{2} \int (\partial_\mu j_\mu) M_c (\partial_\nu j_\nu)] \cdot \\ &\quad \cdot \exp[\tfrac{i}{2} \int j_\mu D^{(T)}_{c,\mu\nu} j_\nu], \end{aligned} \tag{5.23}$$

where a double integration-by-parts has been performed on the $M(x-y)$ dependence of the photon propagator. Because of current conservation expressed by (3.83), and the relation (3.87), it is easy to see that

$$\left[L[\tfrac{1}{i}\tfrac{\delta}{\delta j}],\ \partial_\sigma j_\sigma \right] = 0, \tag{5.24}$$

which means that the longitudinal M_c dependence of the photon propagator is unchanged by the QED interaction. Equation (5.23) may be rewritten in the form

$$NZ^{(M)} = \exp\left\{i\int\bar{\eta}G[\frac{1}{i}\frac{\delta}{\delta j}]\eta\right\} \cdot \exp[-\frac{i}{2}\int(\partial j)M_c(\partial j)] \cdot$$

$$\cdot \exp L[\frac{1}{i}\frac{\delta}{\delta j}] \cdot \exp[\frac{i}{2}\int jD_c^{(T)}j] \tag{5.25}$$

but no further formal manipulation is possible since $[G[\frac{1}{i}\frac{\delta}{\delta j}], \partial j] \neq 0$; that is, each n-point function must be considered separately. The simplest example is

$$D'_{c,\mu\nu}(x-y) = \frac{1}{i}\frac{\delta}{\delta j_\mu(x)}\frac{\delta}{\delta j_\nu(y)} Z^{(M)}\Big|_{\bar{\eta}=\eta=j=0}$$

$$= D'^{(T)}_{c,\mu\nu}(x-y) + \partial^x_\mu\partial^x_\nu M_c(x-y), \tag{5.26}$$

which again states that only the transverse part of the photon propagator is dressed by a conserved-current interaction. Only this portion of the propagator need be renormalized; and indeed, the common multiplicative renormalization is only valid in the Landau gauge, where $M_c = 0$. Under a gauge transformation, $M \to M + \Delta M$, one then has

$$\Delta D'_{c,\mu\nu}(x) = \partial^x_\mu\partial^x_\nu \Delta M_c(x). \tag{5.27}$$

All the M_c, or ΔM_c, dependence can be exhibited for any n-point function. For the electron propagator, one constructs

$$S'_c(x-y) = \frac{1}{i}\frac{\delta}{\delta\bar{\eta}(x)}\frac{\delta}{\delta\eta(y)} Z^{(M)}\Big|_{\eta=\bar{\eta}=j=0}$$

$$= G\left(x,y\middle|\frac{1}{i}\frac{\delta}{\delta j}\right)\exp[-\frac{i}{2}\int(\partial j)M_c(\partial j)]N^{-1} \cdot$$

$$\cdot \exp L[\frac{1}{i}\frac{\delta}{\delta j}]\exp[\frac{i}{2}\int jD_c^{(T)}j]\Big|_o. \tag{5.28}$$

With the aid of (3.89) and the statement (Appendix J) that

$$\left[\frac{\delta}{\delta\Lambda(z)} + \partial_\mu^z \frac{\delta}{\delta A_\mu(z)}\right] G(x,y|A+\partial\Lambda) = 0,$$

simple differentiation yields

$$\left[\partial_\mu^z \frac{\delta}{\delta A_\mu(z)}, G(x,y|A)\right] = -ie[\delta(x-z)-\delta(y-z)]G(x,y|A) \tag{5.29}$$

where the commutator on the LHS of (5.29) is written to denote the functional differentiation of $G(x,y|A)$ only, in case there are other functionals of A present. Interchanging A and $\frac{1}{i}\frac{\delta}{\delta j}$, (5.29) is equivalent to

$$\left[G\left(x,y|\frac{1}{i}\frac{\delta}{\delta j}\right), \partial_\mu^z j_\mu(z)\right] = -e[\delta(x-z)-\delta(y-z)]G\left(x,y|\frac{1}{i}\frac{\delta}{\delta j}\right), \tag{5.30}$$

a result which may also be obtained directly from the differential equations for G, and the easily verified relation

$$\frac{\delta}{\delta A_\mu(z)} G(x,y|A) = ieG(x,z|A)\gamma_\mu G(z,y|A). \tag{5.31}$$

With (5.30), one is in a position to pass the M_c dependence of (5.28) through $G[\frac{1}{i}\frac{\delta}{\delta j}]$ and evaluate the result. It is simplest to define the functional

$$F(\lambda) = G\left(x,y|\frac{1}{i}\frac{\delta}{\delta j}\right)\exp\left[-\frac{i}{2}\lambda\int(\partial j)M_c(\partial j)\right]\mathcal{F}\{j\}, \tag{5.32}$$

and construct the differential equation for $F'(\lambda)$, with the boundary condition

$$F(0) = G\left(x,y\left|\frac{1}{i}\frac{\delta}{\delta j}\right.\right)F\{j\}.$$

One has

$$F'(\lambda) = -\frac{i}{2}\int du\int dv M_c(u,v)\left[G\left(x,y\left|\frac{1}{i}\frac{\delta}{\delta j}\right.\right),\partial j(u)\partial j(v)\right]\cdot$$

$$\cdot\exp\left[-\frac{i}{2}\lambda\int(\partial j)M_c(\partial j)\right]F\{j\}$$

$$-\frac{i}{2}\int(\partial j)M_c(\partial j)\cdot F(\lambda). \qquad (5.33)$$

Using (5.30), the commutator of (5.33) is evaluated as

$$\{e^2[\delta(x-u)-\delta(y-u)]\cdot[\delta(x-v)-\delta(y-v)] - e[\delta(x-u)-\delta(y-u)]\cdot$$

$$\cdot\partial j(v) - e[\delta(x-v)-\delta(y-v)]\cdot\partial j(u)\}\ G\left(x,y\left|\frac{1}{i}\frac{\delta}{\delta j}\right.\right),$$

which yields

$$F'(\lambda) = \{-ie^2[M_c(0)-M_c(x-y)] + ie\int\partial j(w)[M_c(x-w)-M_c(y-w)]$$

$$-\frac{i}{2}\int(\partial j)M_c(\partial j)\}\ F(\lambda),$$

and may be integrated immediately,

$$F(\lambda) = \exp\{-i\lambda e^2[M_c(0)-M_c(x-y)] + i\lambda e\int[\partial j(w)][M_c(x-w) +$$

$$- M_c(y-w)] - \frac{i}{2}\lambda\int(\partial j)M_c(\partial j)\}G\left(x,y\left|\frac{1}{i}\frac{\delta}{\delta j}\right.\right)F\{j\}.$$

Setting $\lambda = 1$ and $j = 0$ wherever possible, (5.28) may be rewritten as

$$S_c'(x-y) = \exp\{-ie^2[M_c(0)-M_c(x-y)]\}G\left(x,y\left|\frac{1}{i}\frac{\delta}{\delta j}\right.\right)N^{-1}\ \cdot$$

$$\cdot \exp L[\frac{1}{i}\frac{\delta}{\delta j}] \cdot \exp[\frac{i}{2}\int jD_c^{(T)}j]\Big|_o, \tag{5.34}$$

or

$$S_c'(x-y)\Big|_{M\neq 0} = \exp\{-ie^2[M_c(0)-M_c(x-y)]\}S_c'(x-y)\Big|_{M=0}, \tag{5.35}$$

which is the desired result. Under a change of gauge, $M \to M + \Delta M$, (5.35) is equivalent to

$$S_c'(x-y)\Big|_{M+\Delta M} = \exp\{-ie^2[\Delta M_c(0)-\Delta M_c(x-y)]\}S_c'(x-y)\Big|_M, \tag{5.36}$$

and it is now necessary to assume that ΔM_c has been sufficiently regularized to be considered finite; for example, if

$$\Delta M_c(x) = \xi\Lambda^2\int d^4k\ e^{ik\cdot x}\ [k^2+\Lambda^2-i\varepsilon]^{-3},$$

where Λ is a regulator mass, and ξ a dimensionless constant, then $M_c(0) = i(\pi^2/2)\xi$, while $\Delta M_c(x)$ vanishes as $|x^2| \to \infty$. Because the wave function renormalization constants are associated with the mass shell singularity of $\tilde{S}_c'(p)$, one expects the configuration-space relation analogous to that of (5.22),

$$\lim_{x\to\infty} S_c'(x) \sim Z_2 \lim_{x\to\infty} S_c(x;m) \tag{5.37}$$

which permits the identification

$$Z_2^{(M+\Delta M)} = Z_2^{(M)}\ e^{-ie^2\Delta M(0)}. \tag{5.38}$$

It must be emphasized that (5.37) or (5.22) require

$\delta > 0$ in (5.19); otherwise, the simple momentum-space pole structure may be masked by a cut. The equivalent statement in this formulation of QED is that the photon is assigned a small mass δ, which is retained until the S-matrix elements and transition probabilities have been computed. From (5.38), one sees that Z_2 is a gauge-dependent constant, in contrast to the gauge-independent Z_3 appearing in the Lehman representation for $\tilde{D}_c'^{(T)}(p)$.

By means of a calculation entirely analogous to that which lead to (5.36), the gauge dependence of any n-point function may be obtained. For example,

$$V_\mu^{(M+\Delta M)}(x,y;z) = \{V_\mu^{(M)}(x,y;z) - ieS_c'^{(M)}(x-y)\frac{\partial}{\partial z_\mu} \cdot$$

$$\cdot [\Delta M(x-z) - \Delta M(y-z)]\} \cdot$$

$$\cdot \exp\{-ie^2[\Delta M(0) - \Delta M(x-y)]\}, \qquad (5.39)$$

and

$$G^{(M+\Delta M)}(x_1y_1,x_2y_2) = G^{(M)}(x_1y_1,x_2y_2) \cdot$$

$$\cdot \exp\{-ie^2[2\Delta M(0) - \Delta\overline{M}]\}, \qquad (5.40)$$

where

$$G(x_1y_1,x_2y_2) \equiv \langle(\psi(x_1)\bar{\psi}(y_1)\psi(x_2)\bar{\psi}(y_2))_+\rangle,$$

and

$$\Delta\overline{M} = \Delta M(x_1-y_1) + \Delta M(x_2-y_2) + \Delta M(x_1-y_2) + \Delta M(x_2-y_1)$$

$$- \Delta M(x_1-x_2) - \Delta M(y_1-y_2).$$

Again, the $\Delta M(0)$ terms are to be interpreted as changes of Z_2 factors, in the manner of (5.38), while the extra configuration space dependence represents the actual changes in these gauge-variant Green's functions. That the S-matrix elements constructed from them are gauge invariant may be understood by observing that if the renormalized $G_R^{(M)} \equiv Z_2^{-2}G^{(M)}$ of (5.40) has mass shell poles in each of the four $\tilde{S}_c'$ propagators associated with each of its external legs, the renormalized $G_R^{(M+\Delta M)}$ will also. The latter may be written as $G_R^{(M)} + \mathcal{R}G_R^{(M)}$, where $\mathcal{R}(x_1,y_1,x_2,y_2)$ denotes the remainder of the expansion of the exponential configuration space-dependent ΔM terms of (5.40),

$$\mathcal{R} = \exp[ie^2\Delta\bar{M}] - 1.$$

If $\tilde{G}_R^{(M)}$ has mass shell poles in every leg, $\widetilde{\mathcal{R}G}_R^{(M)}$ in general will not, and hence will contribute nothing to the S-matrix element, since amputation is an intermediate and crucial step. Thus $G_R^{(M+\Delta M)}$ and $G_R^{(M)}$ may be expected to yield the same S-matrix elements.

The Ward-Takahashi identities are easily derived from the computations of this section, and are illustrated in the simplest way. From (5.14b) one has

$$V_\mu(x,y;z) = iG\left(x,y\middle|\frac{1}{i}\frac{\delta}{\delta j}\right)N^{-1}\exp L\left[\frac{1}{i}\frac{\delta}{\delta j}\right] \cdot$$

$$\cdot\int D_{c,\mu\nu}(z-z')j_\nu(z')\cdot\exp\left[\frac{i}{2}\int jD_c j\right]\Big|_o, \quad (5.41)$$

and upon defining the Klein-Gordon amputated function $V_\mu(x,y;\bar{z})$,

$$V_\mu(x,y;z) = \int D_{c,\mu\nu}(z-z')V_\mu(x,y;\bar{z}'), \quad (5.42)$$

one has

$$V_\mu(x,y;\bar{z}) = iG\left(x,y\Big|\frac{1}{i}\frac{\delta}{\delta j}\right)N^{-1}\exp L[\frac{1}{i}\frac{\delta}{\delta j}]\cdot j_\mu(z)\cdot$$

$$\cdot\exp[\frac{i}{2}\int jD_c j]\Big|_o,$$

from which follows

$$\partial^z_\mu V_\mu(x,y;\bar{z}) = iG\left(x,y\Big|\frac{1}{i}\frac{\delta}{\delta j}\right)\cdot\partial j(z)\cdot N^{-1}\exp L[\frac{1}{i}\frac{\delta}{\delta j}]\cdot$$

$$\cdot\exp[\frac{i}{2}\int jD_c j]\Big|_o, \qquad (5.43)$$

when use has been made of (5.24). Rewriting (5.43) with the aid of (5.30) and the identification (5.13b), one obtains the configuration space statement of the Identity relating vertex and fermion propagator,

$$\partial^z_\mu V_\mu(x,y;\bar{z}) = -ie[\delta(x-z) - \delta(y-z)]S'_c(x-y). \qquad (5.44)$$

Introducing the Fourier representations,

$$S'_c(x) = (2\pi)^{-4}\int d^4Q\ \tilde{S}'_c(Q)\ e^{iQ\cdot x},$$

$$V_\mu(x,y;\bar{z}) = (2\pi)^{-8}\int dp\int dq\int dk\tilde{V}_\mu(q,p;\bar{k})e^{i[q\cdot x+p\cdot y+k\cdot z]},$$

one obtains in momentum space

$$k_\mu\tilde{V}_\mu(q,p;\bar{k}) = e\ \delta(q+p+k)[\tilde{S}'_c(q) - \tilde{S}'_c(q+k)]. \qquad (5.45)$$

If a vertex function Γ_μ (more properly a form factor, but one with the same longitudinal part as the proper vertex function) is introduced according as

$$\tilde{V}_\mu(q,p;\bar{k}) = \delta(q+p+k)\tilde{S}'_c(q)\Gamma_\mu(q,p;k)\tilde{S}'_c(-p),$$

(5.45) may be rewritten as

$$k_\nu \Gamma_\nu(q,q+k;k) = e[\tilde{S}'_c(q+k)^{-1} - \tilde{S}'_c(q)^{-1}], \qquad (5.46)$$

which represents the "integral" form of the W-T Identity.[5] The original "differential" form given by Ward[6] is obtained by calculating $\partial/\partial k_\mu$ of both sides of (5.46), and then setting $k = 0$,

$$\Gamma_\mu(q,q;0) = e \frac{\partial}{\partial q_\mu} \tilde{S}'_c(q)^{-1}. \qquad (5.47)$$

From (5.47) there follows the equality of Z_1 and Z_2, since the former is conventionally defined by the relation

$$\Gamma_\mu(q,q;0)\Big|_{\substack{\text{mass}\\ \text{shell}}} = ie\cdot\gamma_\mu Z_1^{-1}, \qquad (5.48)$$

while near the mass shell, $\gamma\cdot q = im$,

$$\tilde{S}'_c(q) \sim Z_2[m+i\gamma\cdot q]^{-1}.$$

It is clear that a W-T Identity exists, in QED, for every n-point Green's function which displays an external photon index; each such function is then related, by statements of the form (5.46) to other m-point functions, with $n > m$. These identities can be a superb aid to detailed computations.

Notes

1. S. Deser, Phys. Rev. 99, 325 (1955).

2. H. Grotch and D. R. Yennie, Rev. Mod. Phys. 41, 350 (1969).

3. See, for example, any of the texts listed in Reference 4, Chapter 1.

4. This treatment essentially follows that of B. Zumino,

J. Math. Phys. 1, 1 (1960). See also K. Johnson and B. Zumino, Phys. Rev. Letters 3, 351 (1959).

5. Y. Takahashi, Nuovo Cimento 6, 370 (1957).

6. J. C. Ward, Phys. Rev. 78, 182L (1950).

PART II

MODEL APPROXIMATIONS

Chapter 6

PERTURBATION EXPANSIONS

That systematic approximation to any n-point function, in powers of an assumed small coupling constant, is now described with the intention of illustrating perhaps the single most familiar method of approximating Green's functions. The mathematical convergence of the different perturbation series so obtained remains a matter of speculation (and is probably not true); however, these forms possess a definite limiting relationship to phenomenological (e.g., Regge) models, while maintaining a direct relevance to recent experiments which test QED at very small distances.

A. Connectedness and Irreducibility

Examination of the perturbation constructions in field theory has lead to the concept of connected n-point functions, for it is in terms of these quantities that the expansions of Z, or S, are most succinctly given. The situation may be simply illustrated by writing the first few perturbation terms representing the expansion of the 4-point function of a self-interacting boson field $A(x)$, and then inferring the general form

$$M(x_1,x_2,x_3,x_4) = \sum_P \Delta_c'(x_1-x_2)\Delta_c'(x_3-x_4)+M^c(x_1,x_2,x_3,x_4). \tag{6.1}$$

Here, M^c denotes the connected part of M, while the remaining terms on the RHS of (6.1) are the disconnected parts; in the latter, $\sum_P$ denotes the three necessary boson permutations of the coordinate indices x_i. It should be noted that these disconnected terms do not contribute to the corresponding S-matrix element, since they are removed when mass shell amputation is performed on all four coordinates; for higher n-point functions, $n \geq 8$, disconnected terms can contribute to the final amplitude.

One definition of "connected" is that with respect to the separation of any pair of unequal configuration space coordinates, say x and y, as $(x-y)^2 \to \infty$ the connected part of the amplitude will vanish at least as fast as $\exp\{-\mu[(x-y)^2]^{1/2}\}$, where μ denotes the smallest mass in the theory; which is to say that all unequal coordinates of a connected, unrenormalized configuration space amplitude are linked by at least one propagator. Repeating the perturbation construction which lead to (6.1) for higher n-point functions, one finds that the general relation between the complete n-point time-ordered Green's functions and the corresponding connected functions may be expressed by the simple functional statement

$$Z\{j\} = \exp Z^c\{j\}, \tag{6.2}$$

where $Z^c\{j\}$ denotes that generating functional whose differentiation yields all the connected amplitudes. One construction of (6.2) is given in Appendix L. In principle, (6.2) provides a large economy in the amount of effort which must be expended to calculate S-matrix elements in any perturbative, or graphical scheme. Unfortunately, however, formal functional constructions exist for Z, not Z^c.

The concept of irreducibility is essentially that of a refinement, or delineation of the possible forms which connectedness may take. One identifies all the one-particle structure (grouped together in completely dressed propagators) which may link a pair of coordinates, and denotes all remaining structure by the phrase "one-particle irreducible." Of the latter, one isolates all terms in which any two coordinates are linked (in parallel, not in series) by a pair of dressed propagators, and calls the remainder (two-particle irreducible," etc. In terms of momentum space Feynman graphs, one-particle reducible refers to all graphs which may be separated into two distinct parts by cutting a single internal line (propagator), while two-particle reducible identifies those graphs which may be separated into two distinct sections by cutting a pair of lines, etc. This decomposition has been developed especially by Symanzik[1] and by Taylor,[2] and used by them to construct equations of the Bethe-Salpeter type for quite general processes. An elegant way of extracting all one-particle irreducible structure by means of a functional Legendre transformation has been described by Zumino,[3] in connection with an exposition of symmetry and generalized Ward Identities.

B. The Born (Tree Graph) Functional

That approximation in which every n-point function of a specific theory is replaced by its lowest-order perturbative approximation (and pictorially represented by the so-called tree graphs) may be most succinctly defined in terms of an approximation to Z, or to Z^c, from which all virtual-closed-loop structure has been discarded. For simplicity, again consider the self-interaction of a scalar boson field $A(x)$ defined by (3.1) and (3.2) with $n = 3$, and write the functional differential equation for $Z^c\{j\} = \ln Z\{n\}$; from (3.8) this is

$$K_x \frac{1}{i} \frac{\delta}{\delta j(x)} Z^c\{j\} = j(x) + \frac{g}{2} \left\{ \frac{\delta^2 Z^c}{\delta j(x) \cdot \delta j(x)} + \left(\frac{\delta Z^c}{\delta j(x)} \right)^2 \right\}, \tag{6.3}$$

which, together with the boundary condition

$$\delta Z^c/\delta j \Big|_{j=0} = 0,$$

may by simple functional differentiation be used to generate a differential equation (in the x-variable) for every connected n-point function. As in Chapter 5, Section B, every n-point function will be related to an n+1-point function, expressing the complexity (i.e., the virtual structure) of the interaction; and a brief inspection of (6.3) shows that such complexity is due to the presence of the term $\delta^2 Z^c/\delta j(x) \cdot \delta j(x)$, which contains one more functional differentiation of Z^c than appears on the LHS of (6.3). Hence, the approximation which neglects this term,

$$K_x \frac{1}{i} \frac{\delta}{\delta j(x)} Z_B^c\{j\} = j(x) + \frac{g}{2} \left(\frac{\delta Z_B^c}{\delta j(x)}\right)^2 \quad (6.4)$$

may be expected to generate the tree-graph structure appropriate to this interaction. With the notation

$$\tau_B^c(x|j) = \frac{1}{i} \frac{\delta Z_B^c}{\delta j(x)} ,$$

and the corresponding n-point functions given by

$$\tau_B^c(x_1, \cdots x_n) = \tau_B^c(x_1, \cdots x_n|j)\Big|_{j=0}$$

$$= \frac{1}{i} \frac{\delta}{\delta j(x_1)} \cdots \frac{\delta}{\delta j(x_n)} Z_B^c\Big|_{j=0},$$

one functional differentiation of (6.4) yields

$$K_x \tau_B^c(x,y|j) = \delta^4(x-y) - g\tau_B^c(x,y|j)\tau_B^c(x|j), \quad (6.5)$$

and, with $\tau_B^c(x) = 0$, produces $K_x\tau_B^c(x,y) = \delta(x-y)$ or $\tau_B^c(x,y) = \Delta_c(x-y)$. Operation by K_x^{-1} upon (6.5) generates

$$\tau_B^c(x,y|j) = \Delta_c(x-y) - g\int dx'\Delta_c(x-x')\tau_B^c(x',y|j)\tau_B^c(x'|j), \quad (6.6)$$

which has the pictorial representation

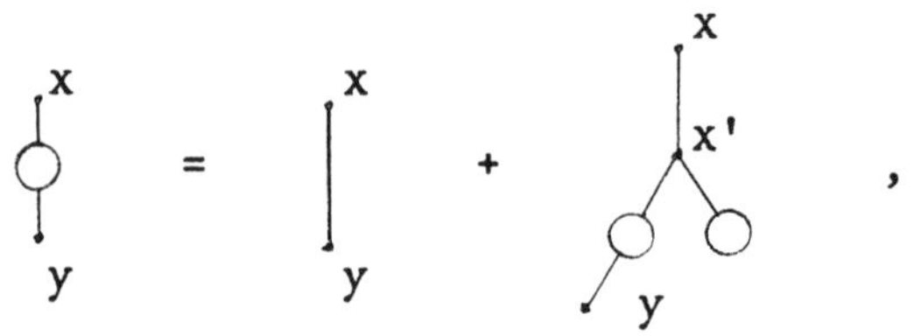

x y = x y + x x' y ,

where each blob $\circ\!\!\!/^{x}$ denotes $\tau_B^c(x|j)$ and is subject to further functional differentiation, producing the higher n-point functions of this approximation.

For the more general interaction (3.2), identical remarks may be made for the solutions to

$$K_x \frac{1}{i} \frac{\delta Z_B^c}{\delta j(x)} = j(x) - \frac{g}{(n-1)!} \left(\frac{1}{i}\right)^{n-1} \left(\frac{\delta Z_B^c}{\delta j(x)}\right)^{n-1}, \tag{6.7}$$

or, still more generally, for the solutions to

$$K_x \frac{1}{i} \frac{\delta Z_B^c}{\delta j(x)} = j(x) + \frac{\partial L'}{\partial A(x)} \left\{\frac{1}{i} \frac{\delta Z_B^c}{\delta j(x)}\right\}. \tag{6.8}$$

Calling $\Phi(x|j) \equiv \frac{1}{i} \frac{\delta Z_B^c}{\delta j(x)}$, one recognizes (6.8) as the equation of motion for a classical field $\Phi(x)$ in the presence of an external source $j(x)$. The question of relevance to quantum field theory of the solution to the classical, nonlinear equation of motion has frequently been raised, and here finds an answer: with an external source, it defines the Born functional, whose n-point functions generate all the tree-graphs.[4]

If one follows factors of $\hbar$ through the preceding analysis, inserting an $\hbar$ in the definition of L (so that $\hbar^{-1}\int d^4x L\{A\}$ is dimensionless) and in the fundamental commutation relations (1.3) (and thereafter a factor of $\hbar$ appears in every propagator), the original equation (3.8) is changed to read

$$K_x \frac{1}{i} \frac{\delta Z}{\delta j(x)} = \hbar j(x) \cdot Z - \frac{g}{(n-1)!} \left(\frac{1}{i} \frac{\delta}{\delta j(x)}\right)^{n-1} Z, \tag{6.9}$$

with formal solution

$$NZ = \exp\left[\frac{i}{\hbar} \int L'\left(\frac{1}{i} \frac{\delta}{\delta j}\right)\right] \cdot \exp\left[\frac{i}{2} \hbar \int j \Delta_c j\right]. \tag{6.10}$$

Changing to the variable $j'(x) = \hbar j(x)$, and rewriting

$Z^c\{j\}$ as $\frac{1}{\hbar} Z^c\{j'\}$, it is easy to see that (6.8) is obtained, in terms of the variable j', with all contributions previously omitted in the passage from (6.3) to (6.8) vanishing in the limit $\hbar \to 0$. In this sense, the tree graphs represent the "classical" part of every amplitude (but with $\mu \to \mu c/\hbar$ unchanged!).

The Born functional may be obtained in the Nambu manner[5] by approximating that functional Fourier transform which expresses the formal generating functional in terms of Feynman's functional (path) integral,

$$Z\{j\} = N' \int d[\phi] \, \exp[iW\{\phi\}], \tag{6.11}$$

a representation easily obtained, as in Chapter 4, from (6.10). In (6.11), $W\{\phi\}$ represents the entire action function, including the source dependence,

$$W = \int d^4x \left(L_o\{\phi\} + L'\{\phi\} + j\phi \right),$$

while N' denotes an appropriate normalization constant. If the exponent of (6.11) is approximated by supposing that the integral is peaked when

$$\phi(x|j) \simeq \phi_o(x|j),$$

$$W\{\phi\} \cong W\{\phi_o\} + \int dz_1 \delta\phi(z_1) \left. \frac{\delta W}{\delta\phi(z_1)} \right|_{\phi=\phi_o} + \frac{1}{2} \int dz_1 \int dz_2 \delta\phi(z_1) \delta\phi(z_2) \left. \frac{\delta^2 W}{\delta\phi(z_1)\delta\phi(z_2)} \right|_{\phi=\phi_o} + \cdots, \tag{6.12}$$

then the condition for an extremum, $\delta W/\delta\phi \big|_{\phi=\phi_o} = 0$, yields the classical equation of motion (6.8), with $\phi_o(x|j) = \Phi(x|j)$. The functional integral of this approximation to (6.11) still contains dependence upon $\phi_o(z|j)$, which may be roughly estimated by retaining only the quadratic terms explicitly written in (6.12), and then permitting the fields $\delta\phi = \phi - \phi_o$ to vary over sufficiently large values to apply the formulae of Appendix K,

$$\ell n\, Z\{j\} = Z^c\{j\} \cong iW\big(\phi_o\{j\}\big) - \frac{1}{2}\,\mathrm{Tr}\,\ell n(1+D_c\Omega), \qquad (6.13)$$

where

$$\langle x|\Omega|y\rangle = \delta(x-y)\cdot\frac{g}{(n-2)!}\,[\phi_o(x|j)]^{n-2}.$$

Because ϕ_o is defined by $\delta W/\delta\phi\big|_{\phi=\phi_o} = 0$, $\delta W/\delta j(x)$ is simply given by considering only the explicit $j(z)$ dependence of W. Hence

$$\frac{\delta W\{\phi_o\}}{\delta j(x)} = \phi_o(x|j) = \frac{1}{i}\,\frac{\delta Z_B^c}{\delta j(x)}\,;$$

and since Z_B^c and W are both to vanish when $j \to 0$, one has $iW\big(\phi_o\{j\}\big) = Z_B^c\{j\}$, thereby re-identifying the first RHS term of (6.13). The remainder, nonzero for $\hbar \neq 0$, represents a particular, model dependent form producing corrections to the tree graph amplitudes, whose singularities remain to be properly redefined (here, by the removal of tadpoles). Exactly analogous considerations may be carried through in the case of several interacting fields. A generalization of the Born functional, with application to chiral symmetry, has recently been given by Wong and Guralnik.[6]

C. Lowest Order Radiative Corrections

A very brief treatment of the simplest radiative corrections to S_c', Γ_μ and $D_{c,\mu\nu}'$ of QED will now be sketched, in order to exhibit the perturbative development directly from the generating functional.

With the aid of (3.61), the formal expression (5.13b) for S_c' may be written as

$$S_c'(x-y) = \exp[-\tfrac{i}{2}\int \frac{\delta}{\delta A} D_c \frac{\delta}{\delta A}]\cdot G(x,y|A)\cdot N^{-1}\exp L[A]\Big|_{A\to 0}. \tag{6.14}$$

A further simplification into the groupings of Appendix I, as in (3.67), is useful in the general case; but to lowest order one may make the replacements $L[A] \to 0$, $N[A] \to 1$, and expand $G(eA)$ to order e^2,

$$S_c'(x-y) \simeq \exp[-\tfrac{i}{2}\int \frac{\delta}{\delta A} D_c \frac{\delta}{\delta A}]\cdot\{S_c(x-y)+(ie)^2\int du\int dv\ \cdot$$

$$\cdot\ S_c(x-u)\gamma\cdot A(u)S_c(u-v)\gamma\cdot A(v)S_c(v-y)\}\Big|_{A\to 0}, \tag{6.15}$$

where terms of this expression odd in A may be dropped. It is clear by inspection that the quadratic A dependence of (6.15) will be replaced by a single photon propagator $-iD_c$, originating in the linear term of the expansion of the exponential operator. A convenient way of describing this is to use the relation

$$\exp[-\tfrac{i}{2}\int \frac{\delta}{\delta A} D_c \frac{\delta}{\delta A}]A_\mu(u)\ \cdots$$

$$= [A_\mu(u)-i\int D_c^{\mu\nu}(u-v)\ \frac{\delta}{\delta A_\nu(v)}\ d^4v]\exp[-\tfrac{i}{2}\int \frac{\delta}{\delta A} D_c \frac{\delta}{\delta A}]\ \cdots,$$

easily derivable from the Baker-Hausdorf forms of Appendix D. The result is

$$S_c'(x-y) \simeq S_c(x-y) + ie^2\int S_c(x-u)\gamma_\mu\ S_c(u-v)\gamma_\nu\ S_c(v-y)\ \cdot$$

$$\cdot\ D_c^{\mu\nu}(u-v), \tag{6.16}$$

which may be given the pictorial representation

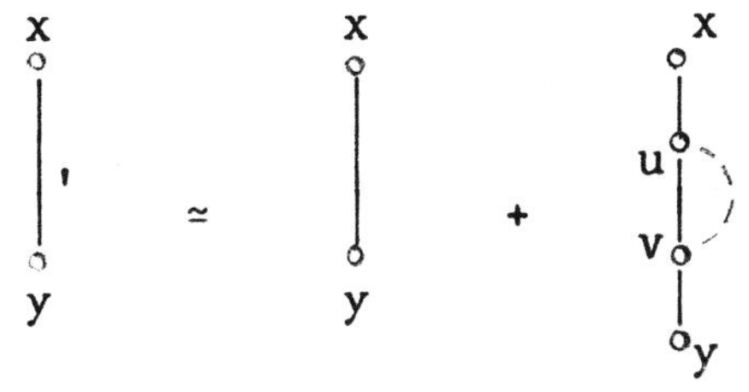

(6.17)

In this configuration-space Feynman graph, one denotes coordinate indices by labeled points joined together by lines which represent fermion or boson propagators; and all internal coordinates, such as u, v, μ, ν are summed upon. Introducing the Fourier representations of Chapter 1, (6.16) may be written as

$$\tilde{S}'_c(p) \simeq \tilde{S}_c(p) + \tilde{S}_c(p)\Sigma_{(2)}(p)\tilde{S}_c(p), \tag{6.18}$$

where $\Sigma_{(2)}(p)$ denotes the e^2 approximation to the fermion self-energy function,

$$\Sigma_{(2)}(p) = \frac{ie^2}{(2\pi)^4}\int d^4k\, \gamma_\mu\, \tilde{S}_c(p-k)\gamma_\nu\, \tilde{D}_c^{\mu\nu}(k), \tag{6.19}$$

and $\tilde{S}'_c(p)$ represents the complete momentum-space propagator, here approximated to include the simplest radiative correction. Momentum-space Feynman graphs are labeled by the momenta carried in each line,

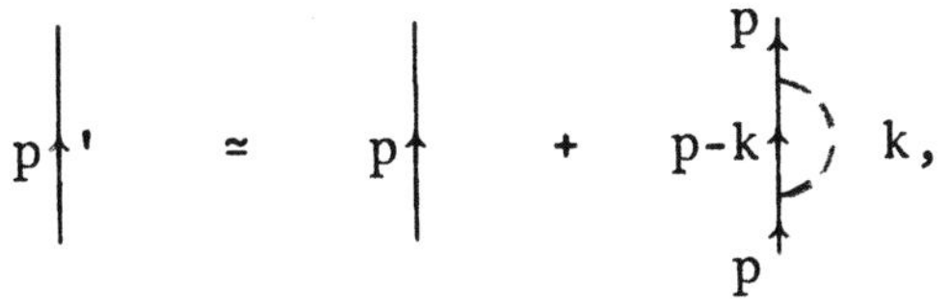

(6.20)

where one conserves four-momenta at every vertex, and sums over all internal coordinates.

Had (6.14) been approximated to include terms of order e^4, one would have found additional contributions to (6.20) of form

$$, \quad (6.21)$$

and it is the first of these, a so-called reducible self-energy term, which when added to (6.20),

$$\tilde{S}_c'(p) \to \tilde{S}_c(p) + \tilde{S}_c(p)\Sigma_{(2)}(p)\tilde{S}_c(p) +$$

$$+ \tilde{S}_c(p)\Sigma_{(2)}(p)\tilde{S}_c(p)\Sigma_{(2)}(p)\tilde{S}_c(p),$$

corresponds to the second iteration of the second-order irreducible self-energy part (6.19) in the relation

$$\tilde{S}_c'(p) \simeq [\tilde{S}_c(p)^{-1} - \Sigma_{(2)}(p)]^{-1}, \quad (6.22a)$$

and suggests the general form for the complete propagator,

$$\tilde{S}_c'(p) = [\tilde{S}_c(p)^{-1} - \Sigma(p)]^{-1}. \quad (6.22b)$$

Here, $\Sigma(p)$ denotes the sum over all irreducible self-energy graphs, $\Sigma = \Sigma_{(2)} + \Sigma_{(4)} + \cdots$, with $\Sigma_{(4)}$ given by the last three graphs of (6.21). The last graph of (6.21) is obtained by keeping the e^2 order expansion of $L[A]$ in (6.14).

The evaluation of (6.19) may be found in standard references, and need not be given here. It may, however, be useful to comment on the general form which follows from (6.22b), and in particular its relation to the Lehman-Källén representation mentioned in Chapter 5. For this it is convenient to consider $\tilde{S}_c'$ and Σ as functions of the variable $\omega = -i\gamma\cdot p$, and rewrite (6.22b) in the form

$$\tilde{S}_c'^{-1}(\omega) = [m_o - \omega - \Sigma(\omega; m_o)] \quad (6.23)$$

where m_o denotes the "bare mass" parameter appearing in the Lagrangian, so that $S_c(p) = (m_o + i\gamma \cdot p)^{-1}$. Adding and subtracting $\Sigma(m;m_o)$ to the RHS of (6.23), where m represents the actual physical mass of the quanta of the interacting fermion field, one obtains

$$\tilde{S}'_c(\omega)^{-1} = m_o - \Sigma(m;m_o) - \omega - [\Sigma(\omega;m_o) - \Sigma(m;m_o)] \tag{6.24}$$

The condition that $\tilde{S}'_c(\omega)$ has a pole at $\omega = m$ defines m in terms of m_o,

$$m = m_o - \Sigma(m;m_o), \tag{6.25}$$

or equivalently, gives m_o in terms of m; either way, this relation has little practical value because of the ultraviolet logarithmic divergence which appears in every order's perturbative approximation to Σ. These relations do provide a useful definition of Z_2, the residue of that pole,

$$Z_2^{-1} = 1 + \left.\frac{\partial \Sigma(\omega;m_o)}{\partial \omega}\right|_{\omega=m}, \tag{6.26}$$

which quantity also has a logarithmic divergence in every order of its perturbation expansion.

The assumption that $\tilde{S}'_c(\omega)$ has a single pole at the physical mass m is frequently made in another way, by the introduction of a mass renormalization counter term in the interaction Lagrangian. The difference between m_o and m is denoted by $\delta m = m_o - m$, and the term $-m_o{:}\bar{\psi}\psi{:}$ of L_o is rewritten as $-m{:}\bar{\psi}\psi{:} - \delta m{:}\bar{\psi}\psi{:}$, with the δm dependence included in the interaction part of the Lagrangian,

$$L' \to ie\bar{\psi}\gamma \cdot A\psi - \delta m{:}\bar{\psi}\psi{:} \tag{6.27}$$

while L_o is now considered as a function of m. It

is easy to see that the exact generating functional, with $\tilde{S}_c(p)$ now containing m, rather than m_o, is unchanged, since

$$N^{-1}Z = \exp[ie\int \frac{\delta}{\delta\eta}\gamma\cdot\frac{\delta}{\delta j}\frac{\delta}{\delta\bar{\eta}}]\cdot\exp[-i\delta m\int :\frac{\delta}{\delta\eta}\frac{\delta}{\delta\bar{\eta}}:]\cdot$$

$$\cdot\exp[i\int\bar{\eta}S_c(m)\eta + \frac{i}{2}\int jD_c j] \tag{6.28}$$

and

$$\exp[-i\delta m\int :\frac{\delta}{\delta\eta}\frac{\delta}{\delta\bar{\eta}}:]\cdot\exp[i\int\bar{\eta}S_c(m)\eta]$$

$$= \exp[i\int\bar{\eta}S_c(m)[1+\delta m S_c(m)]^{-1}\eta] = \exp[i\int\bar{\eta}S_c(m_o)\eta]. \tag{6.29}$$

No infinite phase factor appears in (6.29) because of the use of normal ordering, as in Chapter 3, Section E. Treating δm as a function of e^2, one finds a somewhat more complicated perturbative expansion of the Green's functions following from (6.28), with extra δm dependence entering every fermion self-energy expression corresponding to the shift in nucleon mass, from $\tilde{\Sigma}(\omega;m_o)$ to $\Sigma(\omega;m)$. However, (6.29) guarantees that $\tilde{S}'_c(\omega)$ of (6.23) is unchanged, while δm is again defined by (6.25); in terms of m, considered as a fixed, experimental parameter, this definition of δm reads

$$\delta m = \Sigma(m;m + \delta m). \tag{6.30}$$

Radiative corrections to the vertex function may be obtained from the perturbative expansion of (5.14), written in the form

$$V_\mu(x,y;z) = \exp[-\frac{i}{2}\int\frac{\delta}{\delta A}D_c\frac{\delta}{\delta A}]\cdot iA_\mu(z)G(x,y|A)\cdot N^{-1}\cdot$$

$$\cdot\exp L[A]\Big|_{A\to 0}. \tag{6.31}$$

With the same neglect of closed fermion loops as in the preceding example, expansion of $G(A)$ to order e^3 generates the four contributions most simply represented by the Feynman graphs

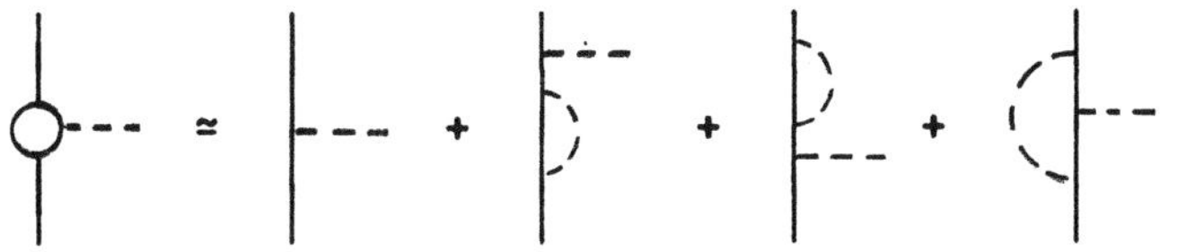

(6.32)

of which the second and third correspond to self-energy insertions in the propagator legs of this 3-point function, while the first and fourth denote the lowest order and first radiative correction to the vertex function Γ_μ, respectively. This is the simplest example of the general structure noted in Chapter 5, Section B, which may be pictorially represented by

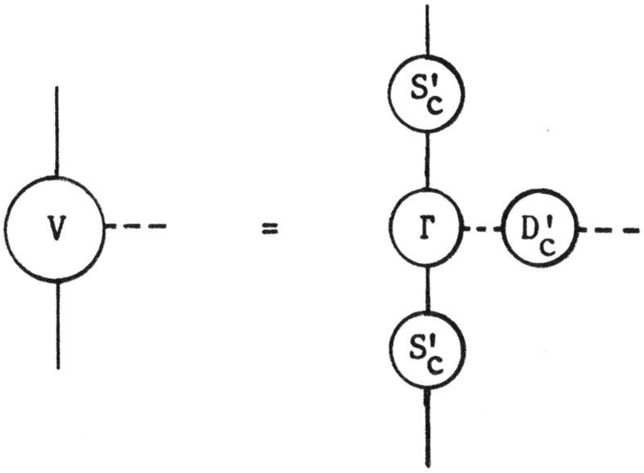

(6.33)

and whose perturbation expansion reproduces (6.32) plus the lowest order closed fermion loop correction to D_c'. Writing $\Gamma_\mu \simeq \Gamma_\mu^{(1)} + \Gamma_\mu^{(3)} + \cdots$, it is now a simple matter, from (6.31), to identify $\Gamma_\mu^{(3)}$ and verify to this order the Ward-Takahashi relation (5.46). Incidentally, starting from the expression (6.32) for V_μ, one may employ the Green's function equation (5.13c) to generate the perturbation expansion of S_c' to order e^4; this involves about as much effort as directly calculating (6.14) to the same order.

Calculation of the lowest order radiative corrections to the photon propagator is of more than passing interest, because of the desirability of obtaining a gauge invariant result directly from the formalism. From the exact generating functional in the absence of fermion sources,

$$NZ\{j\} = \exp[\frac{i}{2} \int j_\mu D_c^{\mu\nu} j_\nu] \cdot \exp[-\frac{i}{2} \int \frac{\delta}{\delta A_\mu} D_c^{\mu\nu} \frac{\delta}{\delta A_\nu}] \cdot$$

$$\cdot \exp L[A], \quad A \equiv \int D_c j,$$

it is easy to see that the formalism automatically produces a photon propagator containing all iterations of the lowest order approximation to $L[A]$, in the sense of (6.22a). Denoting the e^2 order approximation to $L[A]$ by

$$L^{(2)}[A] = \frac{i}{2} \int A_\mu K_{\mu\nu} A_\nu,$$

and the corresponding NZ by $N^{(2)}Z^{(2)}$, there follows from (3.59)

$$N^{(2)}Z^{(2)} = \exp[\frac{i}{2} \int j_\mu D_c^{\mu\nu} j_\nu] \cdot \exp[\frac{i}{2} \int A_\mu [K(1-D_c K)^{-1}]_{\mu\nu} A_\nu] \cdot$$

$$\cdot \exp[\frac{1}{2} \mathrm{Tr}\, \ell n(1-D_c K)^{-1}],$$

or

$$Z^{(2)}\{j\} = \exp[\frac{i}{2} \int j_\mu [D_c(1-KD_c)^{-1}]_{\mu\nu} j_\nu]. \tag{6.34}$$

Thus, the replacement $L \to L^{(2)}$ produces the dressed propagator

$$D'_{c,\mu\nu} = [D_c(1-KD_c)^{-1}]_{\mu\nu}, \tag{6.35}$$

corresponding to the complete iteration of this lowest

order, irreducible self-energy term. From the discussion of Chapter 5, Section C one expects and easily verifies the statement that a gauge invariant $K_{\mu\nu} = \Pi_{\mu\nu}K$ will generate corrections to the gauge invariant part of the photon propagator only; that is, if $D_{c,\mu\nu} = \Pi_{\mu\nu}D_c+\partial_\mu\partial_\nu E_c$ and $D'_{c,\mu\nu} = \Pi_{\mu\nu}D'_c+\partial_\mu\partial_\nu E'_c$, there follow from (6.35) the relations $E' = E$ and

$$D'_c = D_c(1 - KD_c)^{-1}. \tag{6.36}$$

Because of translational invariance, K and D_c and hence D'_c are functions of the difference of their configuration coordinates, with the consequence that (6.36) becomes an algebraic relation between the Fourier transforms of these functions,

$$\tilde{D}'_c(k)^{-1} = \tilde{D}_c(k)^{-1} - \tilde{K}(k). \tag{6.37}$$

In order to show that the proper gauge invariant structure of $K_{\mu\nu}$ does indeed follow from (3.87) and (3.91), one may equate the linear A dependence in the expansion of the latter to $\int K_{\mu\nu}\cdot A_\nu$, or

$$K_{\mu\nu}(x-y) = -e\,\frac{\delta}{\delta A_\nu(y)}\cdot\left\{\lim_{x\to x'} \mathrm{tr}[\gamma_\mu G(x,x'|A)]\cdot\right.$$

$$\left.\cdot\exp\left[-ie\int_{x'}^{x} d\xi_\sigma A_\sigma(\xi)\right]\right\}\Bigg|_{A\to 0}. \tag{6.38}$$

The linear A dependence of the curly bracket of (6.38) is given by

$$\lim_{x\to x'} \mathrm{tr}[\gamma_\mu S_c(x-x')]\left(-ie\int_{x'}^{x} d\xi_\sigma A_\sigma(\xi)\right) \tag{6.39}$$

plus the conventional Feynman graph contribution

$$\lim_{x \to x'} (ie) \int d^4y \, tr[\gamma_\mu S_c(x-y) \gamma \cdot A(y) S_c(y-x')]. \tag{6.40}$$

The latter may be conveniently evaluated by introducing the Fourier representation $A_\nu(y) = \int d^4k \tilde{A}_\nu(k) e^{ik \cdot y}$, and employing the representations (1.11b) and (1.4a), with

$$\Delta_c(z;m^2) = i(2\pi)^{-4} \int_0^\infty da \; e^{-iam^2} \int d^4k \; e^{-iak^2+ik \cdot z}. \tag{6.41}$$

The k-integral is an elementary Gaussian,

$$\int d^4k \; e^{-iak^2+ik \cdot z} = -\frac{\pi^2 i}{a^2} \varepsilon(a) \; e^{iz^2/4a}, \tag{6.42}$$

providing the integral representation for the Bessel function of (1.4c),

$$\Delta_c(z;m^2) = \frac{1}{16\pi^2} \int_0^\infty \frac{da}{a^2} e^{-iam^2+iz^2/4a}. \tag{6.43}$$

Substitution of $S_c(z) = (m - \gamma \cdot \partial^z) \Delta_c(z;m^2)$ into (6.40) leads to

$$\frac{4ie}{(16\pi^2)^2} \int d^4k \tilde{A}_\nu(k) \int_0^\infty \!\! \int \frac{da db}{a^2 b^2} \; e^{-i(a+b)m^2} \int d^4y \; e^{ik \cdot y} \; \cdot$$

$$\cdot \left\{ \delta_{\mu\nu} [m^2 - \frac{(x-y)^2}{4ab}] + \frac{(x-y)_\mu (x-y)_\nu}{2ab} \right\} \cdot$$

$$\cdot \; \exp[\frac{i}{4} (x-y)^2 \left(\frac{1}{a} + \frac{1}{b}\right)]. \tag{6.44}$$

With the aid of the integrals, easily obtained from (6.42),

$$\int d^4y \; \exp[ik \cdot y + \frac{i}{4} \left(\frac{1}{a} + \frac{1}{b}\right) (x-y)^2] =$$

$$= 16\pi^2 i \left(\frac{ab}{a+b}\right)^2 \exp[ik\cdot x - ik^2\left(\frac{ab}{a+b}\right)],$$

and

$$\int d^4y(x-y)_\mu (x-y)_\nu \exp[ik\cdot y + \frac{i}{4}\left(\frac{1}{a}+\frac{1}{b}\right)(x-y)^2]$$

$$= 16\pi^2 i\left(\frac{ab}{a+b}\right)^3 \{2i\delta_{\mu\nu} + 2k_\mu k_\nu \left(\frac{ab}{a+b}\right)\} \cdot$$

$$\cdot \exp[ik\cdot x - ik^2\left(\frac{ab}{a+b}\right)]$$

one may rewrite (6.44) in the form

$$-\frac{e}{4\pi^2}\int d^4k\tilde{A}_\nu(k)e^{ik\cdot x}\int\int_0^\infty \frac{dadb}{(a+b)^2}\exp[-im^2(a+b)-ik^2\left(\frac{ab}{a+b}\right)]\cdot$$

$$\cdot\{\delta_{\mu\nu}[m^2+k^2\cdot\frac{ab}{(a+b)^2} - \frac{i}{(a+b)}] + \frac{2ab}{(a+b)^2}(k_\mu k_\nu - k^2\delta_{\mu\nu})\}. \tag{6.45}$$

The familiar logarithmic and quadratic ultraviolet divergences of (6.40) now appear in the lower limit dependence of the parametric a,b variables, with (6.45) separated into explicit gauge variant and gauge invariant combinations. A final change of variable, $a = xu$, $b = yu$,

$$\int\int_0^\infty dadb = \int_0^\infty udu\int_0^1 dx\int_0^1 dy\ \delta(x+y-1),$$

converts this into the somewhat more convenient form

$$-\frac{e}{4\pi^2}\int d^4k\tilde{A}_\nu(k)e^{ik\cdot x}\int_0^\infty\frac{du}{u}\int_0^1 dy\ \exp[-iu\left(m^2+y(1-y)k^2\right)]\ \cdot$$

$$\cdot\{\delta_{\mu\nu}[m^2+y(1-y)k^2-\frac{i}{u}] + 2y(1-y)(k_\mu k_\nu - k^2\delta_{\mu\nu})\}. \qquad (6.46)$$

To this result must now be added the gauge variant contribution of (6.39), whose evaluation is made simplest by the adoption of a straight line path between the points x and x', $\xi_\sigma = \lambda x_\sigma + (1-\lambda)x'_\sigma$, and

$$\int_{x'}^{x} d\xi_\sigma A_\sigma(\xi) = \varepsilon_\sigma \int_0^1 d\lambda A_\sigma(x'+\lambda\varepsilon) = \varepsilon_\sigma \int_0^1 d\lambda A_\sigma(x-\lambda\varepsilon),$$

with $\varepsilon_\sigma = x_\sigma - x'_\sigma$. The quantity $tr[\gamma_\mu S_c(\varepsilon)]$ is singular as $\varepsilon_\mu/\varepsilon^2\cdot\varepsilon^2$, and hence those contributions entering from

$$\left(-ie\int_{x'}^{x} d\xi_\sigma A_\sigma(\xi)\right)$$

will be

$$-ie\{\varepsilon_\sigma A_\sigma(x) + \frac{1}{6}\varepsilon_\sigma\left(\varepsilon\cdot\frac{\partial}{\partial x}\right)^2 A_\sigma(x)\},$$

with overall odd ε dependence discarded in this symmetric limit. Introducing Fourier representations of $A_\sigma(x)$ and Δ_c, one obtains for (6.39)

$$-\frac{ie}{(2\pi)^4}\int d^4k\tilde{A}_\sigma(k)e^{ik\cdot x}\int\frac{d^4p}{p^2+m^2}\, tr[\gamma_\mu(m-i\gamma\cdot p)]\,\cdot$$

$$\cdot\ \{\varepsilon_\sigma - \frac{1}{6}\varepsilon_\sigma(\varepsilon\cdot k)^2\}e^{ip\cdot\varepsilon}$$

$$= \frac{4ie}{(2\pi)^4}\int d^4k\tilde{A}_\sigma(k)e^{ik\cdot x}\int\frac{d^4p}{p^2+m^2}\, p_\mu\frac{\partial}{\partial p_\sigma}\{1+\frac{1}{6}\left(k\cdot\frac{\partial}{\partial p}\right)^2\}e^{ip\cdot\varepsilon}.$$

(6.47)

An integration by parts before taking the limit $\varepsilon \to 0$ leads to

$$- \frac{4ie}{(2\pi)^4} \int d^4k \tilde{A}_\sigma(k) e^{ik\cdot x} \int d^4p e^{ip\cdot\varepsilon} \left\{1 + \frac{1}{6}\left(k\cdot\frac{\partial}{\partial p}\right)^2\right\} \frac{\partial}{\partial p_\sigma} \cdot$$

$$\cdot \left(\frac{p_\mu}{p^2+m^2}\right), \tag{6.48}$$

and the limit $\varepsilon \to 0$ now converts (6.48) into the divergent surface integral, or equivalent volume integral obtained by carrying through the $\partial/\partial p$ differentiations of (6.48) with the symmetric integration replacements

$$\int d^4p \; p_\mu p_\nu f(p^2) \to \frac{1}{4} \delta_{\mu\nu} \int d^4p \cdot p^2 f(p^2),$$

$$\int d^4p \cdot p_\mu p_\nu p_\lambda p_\sigma f(p^2) \to \frac{1}{24} \int d^4p \cdot p^2 p^2 f(p^2) \cdot$$

$$\cdot \{\delta_{\mu\nu}\delta_{\lambda\sigma} + \delta_{\mu\lambda}\delta_{\nu\sigma} + \delta_{\mu\sigma}\delta_{\nu\lambda}\},$$

$$- \frac{e}{4\pi^2} \int d^4k \tilde{A}_\sigma(k) e^{ik\cdot x} \left\{ \delta_{\mu\sigma} \left(\frac{i}{\pi^2}\right) \int d^4p \; \frac{m^2 + \frac{1}{2} p^2}{(m^2+p^2)^2} \right.$$

$$\left. - \frac{1}{3} k_\nu k_\lambda [\delta_{\mu\nu}\delta_{\lambda\sigma} + \delta_{\mu\lambda}\delta_{\nu\sigma} + \delta_{\mu\sigma}\delta_{\nu\lambda}] \left(\frac{i}{\pi^2}\right) \int d^4p \cdot \frac{m^4}{(m^2+p^2)^4} \right\}. \tag{6.49}$$

That divergent term independent of k in the curly bracket of (6.49) may be written as

$$\left(\frac{i}{\pi^2}\right)\delta_{\mu\sigma}\cdot i^2\int_0^\infty udu\ e^{-ium^2}\int d^4p\ e^{-iup^2}[m^2 + \frac{1}{2}\ p^2]$$

$$= -i\delta_{\mu\sigma}\int_0^\infty du\ \frac{\partial}{\partial u}\ [\frac{1}{u}\ e^{-ium^2}] = \delta_{\mu\sigma}\left(m^2 + \frac{i}{u}\bigg|_{u\to 0}\right), \tag{6.50}$$

while the finite term, quadratic in k, is simply

$$\frac{1}{18}\ [\delta_{\mu\sigma}k^2 + 2k_\mu k_\sigma]. \tag{6.51}$$

The gauge variant portion of the corresponding part of (6.46) may be written as

$$i\delta_{\mu\sigma}\int_0^1 dy\int_0^\infty du\ \frac{\partial}{\partial u}\ [\frac{1}{u}\ e^{-iu(m^2+y(1-y)k^2)}]$$

$$= -\delta_{\mu\sigma}(m^2 + \frac{i}{u}\bigg|_{u\to 0} + \frac{1}{6}\ k^2), \tag{6.52}$$

and one now verifies that the sum of (6.50), (6.51) and (6.52) is the gauge invariant combination

$$\frac{1}{9}\ [k_\mu k_\sigma - k^2\delta_{\mu\sigma}], \tag{6.53}$$

with the gauge variant quadratic divergence removed. In Feynman graph language, the k-independent part of (6.48) exactly cancels the k-independent quadratic divergence of this simplest closed fermion loop; but there remains in the once-subtracted Feynman integral a finite, gauge variant term quadratic in k, here removed by the corresponding portion of (6.48). Because the line integral

$$\left(\int_{x'}^{x} d\xi_\sigma A_\sigma(\xi)\right)^n$$

vanishes at least as fast as ε^n, similar considerations hold for the closed fermion loop with four photon legs, $K_{\mu\nu\lambda\sigma}$, where the exponential line integral factor generates finite contributions needed for gauge invariance. Higher n-point amplitudes of this photon-photon variety (excluding vertex and self-energy insertions) are gauge invariant and finite, without the need for the line integral.

Grouping the contribution of (6.53) together with the remaining gauge invariant part of (6.46), one obtains

$$\tilde{K}_{\mu\nu}(k) = [\delta_{\mu\nu} - \frac{k_\mu k_\nu}{k^2}]\ K(k^2),$$

with $K(k^2) = -k^2\Pi(k^2)$ and

$$\Pi(k^2) = +\frac{e^2}{2\pi^2}\int_0^1 dy\cdot y(1-y)\left\{\int_0^\infty \frac{du}{u}\, e^{-iu[m^2+y(1-y)k^2]} + \frac{1}{3}\right\}, \tag{6.54}$$

and verifies that (6.54) leads to a dressed photon propagator, as in (6.37), of zero mass, since $K(k^2)$ vanishes with k^2. Near the mass shell, $k^2 \sim 0$, (6.37) provides an expression for

$$\tilde{D}'_c(k) \sim Z_3/k^2 + \cdots,$$

from which one may read off the photon's wave function renormalization constant Z_3^{-1} of this approximation, here given by the real, positive, logarithmically divergent quantity $1+\Pi(0)$. The inverse of this expression, for Z_3, contains all powers of e^2 and is somewhat deceptive, since the only input into the calculation has been the e^2 dependence of $L[A]$; only the e^2 portion of Z_3,

$$Z_3^{(2)} = 1-\Pi(0) = 1-\frac{e^2}{12\pi^2}\left\{\int_0^\infty \frac{ds}{s}\, e^{-sm^2} + \frac{1}{3}\right\} \tag{6.55}$$

will be unchanged when the higher-order terms of $L[A]$ are included. It is not known whether the divergence of this gauge invariant quantity reflects the inadequacy of the perturbation expansion or of the underlying field theory, but there is hope in field theoretic circles that the first of these possibilities is true.[7]

The total, or effective, current appearing as a result of an external field $\tilde{A}^{ext}_\mu(k)$ is given by the sum of $\tilde{j}^{ext}_\mu(k) = k^2\tilde{A}^{ext}_\mu(k)$ and the induced current $<\tilde{j}_\mu(k)>_{A^{ext}}$, in which context--now that $\tilde{K}_{\mu\nu}(k)$ has been shown to have the proper gauge structure-- $\tilde{A}^{ext}_\mu$ may be taken to satisfy the Lorentz condition

$$k_\mu \tilde{A}^{ext}_\mu(k) = 0.$$

Hence

$$<\tilde{j}^{eff}_\mu(k)> = [1-\Pi(k^2)]\ \tilde{j}^{ext}_\mu(k)$$

or

$$e<\tilde{j}^{eff}_\mu(k)> = [1-\Pi(0)]e\tilde{j}^{ext}_\mu(k)-[\Pi(k^2)-\Pi(0)]e\tilde{j}^{ext}_\mu(k). \tag{6.56}$$

Since $\tilde{j}^{ext}_\mu(k)$ is proportional to the electric charge, it may be written as $e\tilde{J}^{ext}_\mu(k)$, where J^{ext}_μ is independent of e. With (6.55), to this order one renormalizes the original coupling by writing

$$e<\tilde{j}^{eff}_\mu(k) = Z_3 e^2\tilde{J}^{ext}_\mu(k)-[\Pi(k^2)-\Pi(0)]e\tilde{j}^{ext}_\mu(k), \tag{6.57}$$

and requiring the classical limit of QED: at distances $>> m^{-1}$ ($= \hbar/mc$, the electron Compton wavelength) the renormalized coupling must be identified with the classical charge e_R. In momentum space this means

$$\lim_{k\to 0} e \langle \tilde{j}_\mu^{eff}(k) \rangle = e_R^2\, \tilde{J}_\mu^{ext}(0), \qquad (6.58)$$

and comparison with (6.57) provides the identification $e_R^2 = e^2 Z_3$; to this order, (6.57) may be rewritten as

$$e\langle \tilde{j}_\mu^{eff}(k) \rangle = e_R^2\, \tilde{J}_\mu^{ext}(k) - [\Pi(k^2) - \Pi(0)]\, e\tilde{j}_\mu^{ext}(k). \qquad (6.59)$$

The Dyson-Ward proof that such absorption of Z_3 factors in the redefinition of physical charge may be performed to all orders in e^{2n} is far beyond the scope of these remarks, and may be found elsewhere;[8] but it is difficult to avoid mentioning the lovely, qualitative argument due to Schwinger[9] which succinctly describes the essential physics: Any "bare" charge e surrounds itself with an induced, or vacuum polarization charge of amount $(+\delta e)+(-\delta e)$, with the total induced charge zero. The $+\delta e$ charge can escape to infinity, leaving $-\delta e$ behind at a distance $\sim m^{-1}$ from e. If one observes the charge at distances $\gg m^{-1}$, one measures $e-\delta e$, and calls this the physical, or renormalized charge e_R; but if one probes at distances smaller than m^{-1}, one measures e, not e_R, and $e > e_R$.

In the difference $\pi(k^2)-\pi(0)$ of (6.59),

$$\Pi(k^2)-\Pi(0) = \frac{e^2}{2\pi^2}\int_0^1 dy\cdot y(1-y)\int_0^\infty \frac{du}{u}\, e^{-ium^2} \cdot$$

$$\cdot\, [e^{-iuy(1-y)k^2} - 1], \qquad (6.60)$$

it is convenient to replace the square bracket of (6.60) by

$$\int_0^1 d\lambda\, \frac{\partial}{\partial\lambda}\, e^{-i\lambda uk^2 y(1-y)},$$

carry out the $\partial/\partial\lambda$ differentiation, then the $\int_0^\infty du$

integration, and finally the $\int_0^1 d\lambda$ integration, in order to obtain the somewhat more familiar form

$$-\frac{2\alpha}{\pi}\int_0^1 dy\cdot y(1-y)\ \ell n[1+y(1-y)k^2/m^2]. \qquad (6.61)$$

Discussion of the applications of (6.61) and other radiative corrections considered in this section may be found in standard texts, along with "rules" for writing down arbitrarily complicated Feynman graphs. Such prescriptions usually conclude with the instruction to sum over all topologically distinct graphs. All such rules, including the last one, are quite readily obtained from the perturbative expansion of the generating functional.

D. Renormalization Procedures

The manner in which renormalization constants are removed from the generalized reduction formula of the S-matrix, (2.25), will now be briefly sketched, continuing with the example of (massive photon) QED. The connected generating functional

$$Z^{(c)}\{j_\mu,\eta,\bar{\eta}\} = \ell n\ Z\{j_\mu,\eta,\bar{\eta}\}$$

may be put into the form

$$Z^{(c)} = \frac{i}{2}\int j_\mu D'_{c,\mu\nu} j_\nu + i\int \bar{\eta} S'_c \eta + i\int \bar{\eta} V_\mu j_\mu \eta + \frac{i}{2}\int \bar{\eta}\bar{\eta} M^{(c)} \eta\eta +$$

$$+ \frac{i}{2}\int \bar{\eta} C^{(c)}_{\mu\nu} j_\mu j_\nu \eta + \cdots \qquad (6.62)$$

where D'_c, S'_c denote the completely dressed, unrenormalized propagators, and V_μ, $M^{(c)}$, $C^{(c)}_{\mu\nu}$, are the connected vertex, lepton-lepton, and Compton scattering amplitudes, respectively. The latter three functions contain dressed propagators of the appropriate type on each of their legs, which may be represented pictorially by

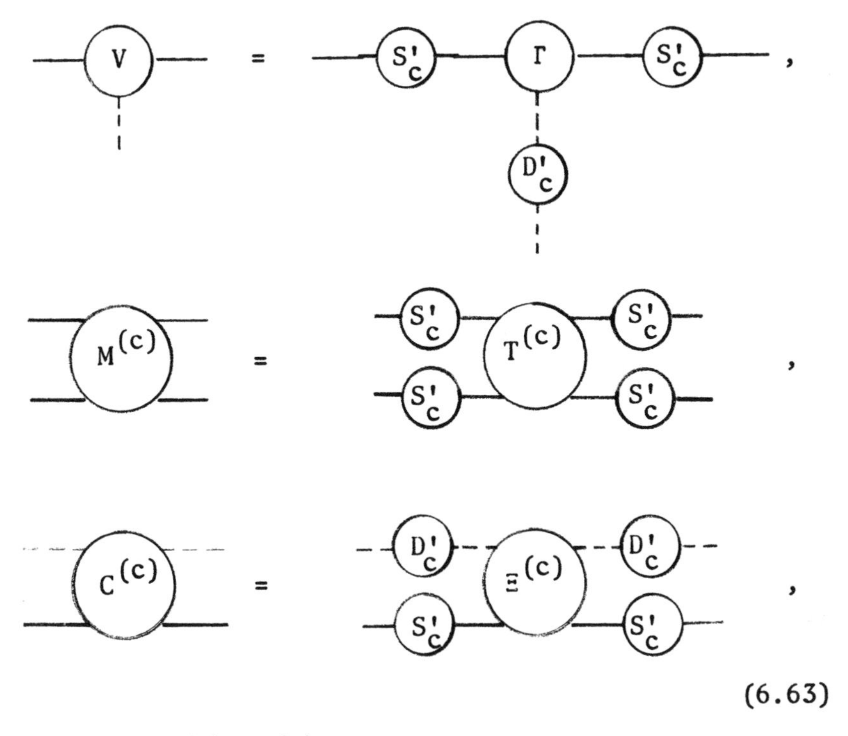

(6.63)

where Γ_μ, $T^{(c)}$, $\Xi^{(c)}_{\mu\nu}$, represent the completely amputated functions.

It is convenient to define renormalized sources,

$$j^\mu_R = \sqrt{Z_3}\, j^\mu, \quad \eta_R = \sqrt{Z_2}\, \eta, \quad \bar{\eta}_R = \sqrt{Z_2}\, \bar{\eta},$$

to remove the $Z^{-1/2}$ factors from the Wick-ordered products of (2.25),

$$\frac{S}{\langle S \rangle} = :\exp[A_{IN}K \frac{\delta}{\delta j_R} + \bar{\psi}_{IN} \vec{D} \frac{\delta}{\delta \bar{\eta}_R} - \frac{\delta}{\delta \eta_R} \overleftarrow{D}\, \psi_{IN}]: \cdot Z\left\{\frac{j^\mu_R}{\sqrt{Z_3}}, \frac{\eta_R}{\sqrt{Z_2}}, \frac{\bar{\eta}_R}{\sqrt{Z_2}}\right\}, \quad (6.64)$$

with the Z factors now appearing in (6.62). With one minor qualification, these factors are just those required to convert the bare charge e to the renormalized charge $e_R = \left((Z_2/Z_1)\sqrt{Z_3}\,\right)e$ in every perturbative approximation to each S-matrix element. That qualification lies in the replacement $Z_3^{-1}D'_{c,\mu\nu}$ by $D'^{(R)}_{c,\mu\nu}$, a statement strictly true in the Lorentz (or Landau) gauge, where $D'_{c,\mu\nu} = \Pi_{\mu\nu}D'_c$, for only the transverse part of the photon propagator is dressed by the current-conserving interaction. However, as noted in Chapter 5, Section C, the gauge dependent part of the photon propagator does not contribute to S-matrix elements, which property remains true even when improperly renormalized.

In terms of renormalized sources, the first and second terms of $Z^{(c)}$ are

$$\frac{i}{2}\int j_R^{\mu}\, D'^{(R)}_{c,\mu\nu}\, j_R^{\nu} + i\int \bar{\eta}_R\, S'^{(R)}_c\, \eta_R,$$

while the third may be written as

$$ie_R\int \bar{\eta}_R\, S'^{(R)}_c\, \Gamma^{(R)}_{\mu}\, S'^{(R)}_c\, \eta_R\cdot D'^{(R)}_{c,\mu\nu}\, j_R^{\nu},$$

where the renormalized vertex $\Gamma^{(R)}_{\mu}$ is defined as $Z_1^{-1}\Gamma_{\mu}$, with Z_1 given by (5.48) and equal to Z_2 in QED. Similarly, the lepton-lepton scattering amplitude of (6.62) becomes

$$\frac{i}{2}\int\left(\bar{\eta}_R S'^{(R)}_c\right)\left(\bar{\eta}_R S'^{(R)}_c\right)\cdot Z_2^2 T_c\cdot\left(S'^{(R)}_c\eta_R\right)\left(S'^{(R)}_c\eta_R\right),$$

while the perturbation expansion of T_c may be most conveniently defined by "fleshing in" those skeleton graphs obtained from the linkage of lowest order vertex and propagator,

Chapter 7

SOLUBLE MODELS

Two exactly soluble models of quantum field theory have been constructed for the special case of zero-bare-mass in two-dimensional space-time. While the physical content of such theories turns out to be rather small, they have provided examples of exact, finite Green's functions, and have lead to some interesting speculations concerning the definition of current operators. The emphasis here will not be on the latter, however, but simply on the models as examples of complete solubility. An example of the "numerical" models of Caianiello is then briefly treated, and a comment made on the partially soluble Lee model, at the end of the chapter.

A. Two-dimensional Electrodynamics

This model was put forth by Schwinger[1] for the purpose of demonstrating that gauge invariance need not require a massless photon, a condition and result one might be tempted to associate on the basis of the previous perturbation calculation. To obtain the generating functional (3.74), with the modification (3.91), one begins with the lepton, potential theory Green's function satisfying

$$\sum_{\mu} \gamma_{\mu}(\partial - ieA)_{\mu} G(x,y|A) = \delta^{(2)}(x-y), \tag{7.1}$$

where the restriction to zero mass is necessary for solubility. Although there is no bare mass in this theory, it should be noted that the coupling constant e has the dimensions of mass, in two-dimensional space-time. The index μ is either 1 or 4; and it is sometimes convenient to write $\gamma\cdot\partial = \gamma_1\partial_1 + \gamma_4\partial_4$, choosing $\gamma_1 = \sigma_1$ and $\gamma_4 = \sigma_3$, the hermitian Pauli matrices with real elements. One expects a solution of (7.1) to be given in terms of the corresponding free-field propagator,

$$S_c(x) = \frac{i}{2\pi} \frac{\gamma\cdot x}{x^2+i\varepsilon}, \tag{7.2}$$

which may be obtained by elementary Gaussian integration from the relations $S_c = -(\gamma\cdot\partial)D_c$ and

$$D_c(x) = (2\pi)^{-2}\int d^2k[k^2-i\varepsilon]^{-1}\, e^{ik\cdot x}$$

$$= \frac{i}{4\pi}\int_0^\infty dt\cdot t^{-1}\, e^{ix^2/4t}.$$

The latter representation is logarithmically divergent, but the difference $D_c(x)-D_c(x_o)$ is finite, as is the $S_c(x)$ of (7.2).

Because of the restriction to two dimensions, it is possible to find a solution based upon the ansatz

$$G(x,y|A) = \exp[ie(\phi(x)-\phi(y))]\cdot S_c(x-y), \tag{7.3}$$

where $\gamma\cdot\partial\phi(x) = \gamma\cdot A(x)$. This in turn means that $(\gamma\cdot\partial)^2\phi = \partial^2\phi = (\gamma\cdot\partial)\gamma\cdot A$, implying the causal solution

$$\phi(x) = -\int D_c(x-x')(\gamma\cdot\partial')(\gamma\cdot A(x')), \tag{7.4}$$

where we do not bother to replace $D_c(x-x')$ by $D_c(x-x')-D_c(x_o-x')$, since such subtraction is not necessary for the finite exponential factor of (7.3). It is clear from the rules of differentiation of exponentials, that (7.3) and (7.4) provide a solution only if $[\partial_\mu\phi,\phi] = 0$, which will be true if $[\gamma_\mu\gamma_\nu,\gamma_\lambda\gamma_\sigma] = 0$. This condition, false in four dimensions, is true here since there are but two matrices, γ_1 and γ_4. Hence the product $\gamma_\mu\gamma_\nu$ may be either 1 or $\gamma_1\gamma_4 = -\gamma_4\gamma_1$, each of which commutes with itself and the other matrix. Thus, in two dimensions, (7.3) with (7.4) represents the exact, causal solution to (7.1).

It is now possible to calculate exactly the closed-loop functional $L[A]$ defined by (3.86) and (3.91),

$$L[A] = -i\int d^2x A_\mu(x)\int_0^e de' \lim_{x\leftrightarrow x'} tr[\gamma_\mu G(x,x'|e'A)] \cdot$$

$$\cdot \exp[-ie'\int_{x'}^{x} d\xi_\sigma A_\sigma(\xi)]. \qquad (7.5)$$

The space-like nature of the limit is guaranteed by first choosing $x_0' = x_0$, and calculating (7.5) in the limit as $x_1' \leftrightarrow x_1$. From (7.2) it follows that one need retain no higher than linear terms in $\varepsilon_1 \equiv x_1 - x_1'$ arising from expansion of the exponential factor,

$$L[A] = \frac{1}{2\pi}\int d^2x A_\mu(x)\int_0^e de' \lim_{\varepsilon_1\to 0} tr\Bigg[\gamma_\mu \cdot$$

$$\cdot \{1+ie'\varepsilon_1[\partial_1\phi(x)-A_1(x)]\}\frac{\gamma\cdot\varepsilon_1}{\varepsilon_1^2}\Bigg]. \qquad (7.6)$$

Because the limit is to be taken symmetrically, all that remains is the finite quantity

$$L[A] = \frac{ie^2}{4\pi}\int d^2x A_\mu(x)\; tr[\gamma_\mu\big(\partial_1\phi(x)-A_1(x)\big)\gamma_1]. \qquad (7.7)$$

From the defining equation for ϕ, it follows that

$$tr[\gamma_\mu(\partial_1\phi - A_1)\gamma_1] = -2[A_\mu - \frac{1}{2}\partial_\mu tr\phi],$$

while

$$\frac{1}{2} tr\phi(x) = -\int D_c(x-x')\partial_\sigma' A_\sigma(x');$$

and there then results the exact, and simple form

$$L[A] = -\frac{ie^2}{2\pi}\int d^2x A_\mu(x) A_\mu^T(x)$$

$$= -\frac{ie^2}{2\pi}\int d^2x A_\mu(x)\Pi_{\mu\nu}(\partial)A_\nu(x), \tag{7.8}$$

where A_μ^T denotes the transverse part of A_μ,

$$A_\mu^T(x) = A_\mu(x) + \partial_\mu^x \int D_c(x-x')\partial_\sigma' A_\sigma(x'),$$

with $\partial_\mu A_\mu^T \equiv 0$. Assuming the A_μ defined so that the partial integrations common to four dimensions are permissible, one finds $\langle j_\mu \rangle_A \sim A_\mu^T$, a Lorentz covariant and gauge invariant result. Because $L[A]$ is a quadratic functional of A, while $G[A]$ is given as the exponential of a linear form in A, all functional operations involved in the calculation of every n-point function may be performed exactly.

It should be noted that the exponential line integral used in the definition of $L[A]$ is an essential ingredient in producing both a gauge invariant and Lorentz covariant result. Without this factor, a covariant result is not automatically obtained, since $\langle j_\mu \rangle$ will then be proportional to $\varepsilon_\mu \varepsilon_\nu / \varepsilon^2$, which quantity depends upon the way ε vanishes. In the literature,[2] this difficulty has been treated by adopting a time-like as well as a space-like limiting procedure, replacing this quantity by one proportional to $\delta_{\mu\nu}$, a procedure which may be made to yield a properly gauge invariant result, but brings into question the concept of a local Lagrangian.

The generating functional (3.74) may now be written down in terms of the solutions (7.3) and (7.8); for simplicity, subsequent calculations are performed in the Lorentz gauge, with $D_{c,\mu\nu} = \Pi_{\mu\nu} D_c$. In the absence of fermion sources, $\eta = \bar{\eta} = 0$, the functional takes the form

$$Z\{j\} = N^{-1} \cdot \exp[\frac{i}{2}\int j \cdot \Pi D_c \cdot j] \cdot \exp[-\frac{i}{2}\int \frac{\delta}{\delta A} \cdot \Pi D_c \frac{\delta}{\delta A}] \cdot$$

$$\cdot \exp[\frac{i}{2}\int A \cdot \Pi[-\frac{e^2}{\pi}] \cdot A], \tag{7.9}$$

where $[-e^2/\pi]$ denotes the unit operator, in configuration and $\mu\nu$ space, multiplied by the numerical constant $-e^2/\pi$; again, $A_\mu \equiv \int D_c^{\mu\nu} j_\nu$. With (3.59), this is evaluated as

$$Z\{j\} = \exp\left[\frac{i}{2}\int j \cdot \Pi D_c j + \frac{i}{2}\int A \cdot \Pi[-\frac{e^2}{\pi}]\left(1 - D_c[-\frac{e^2}{\pi}]\right)^{-1} A\right], \tag{7.10}$$

where N has been identified, in the limit of vanishing source, as

$$N = \exp\left[\frac{1}{2} \operatorname{Tr} \ln\left(1 - D_c[-\frac{e^2}{\pi}]\right)^{-1}\right]. \tag{7.11}$$

Combining terms in the exponent of (7.10), one finds

$$Z\{j\} = \exp[\frac{i}{2}\int j \cdot \Pi \Delta_c \cdot j], \tag{7.12}$$

with

$$\tilde{\Delta}_c(p) = [p^2 + \frac{e^2}{\pi} - i\varepsilon]^{-1}. \tag{7.13}$$

In this exact, gauge invariant result of two dimensions, the photon has acquired a mass, $e/\sqrt{\pi}$. The theory, however, is essentially a free-field one, since there are no connected photon-photon n-point functions for $n > 2$. Incidentally, upon supplying a momentum (Λ) and a configuration space (LT) cut off, the "vacuum-to-vacuum" amplitude of (7.11) may be shown to be an infinite phase factor,

$$N \sim \exp[-\frac{ie^2}{8\pi^2}(LT)\int_0^1 du \, \ell n\left(\frac{\Lambda^2}{e^2/\pi}u\right)].$$

The corresponding calculation of the simplest fermion function, S_c', is also immediate. With (6.14), one has

$$S_c'(x-y) = \exp[-\frac{i}{2}\int \frac{\delta}{\delta A}\Pi D_c \frac{\delta}{\delta A}]\cdot G(x,y|A)\cdot N^{-1}\cdot$$

$$\cdot \exp\left(\frac{i}{2}\int A\cdot\Pi[-\frac{e^2}{\pi}]\cdot A\right)\Big|_{A=0}, \tag{7.14}$$

with $G[A]$ given by (7.3). Again employing (3.59), and the identification (7.11), one obtains

$$S_c'(x-y) = \exp[\frac{ie^2}{2}\int f_\mu \Delta_c \cdot \Pi_{\mu\nu}\cdot f_\nu] \cdot S_c(x-y), \tag{7.15}$$

with

$$f_\mu(z) = \gamma\cdot\partial^z[D_c(x-z)-D_c(y-z)]\gamma_\mu$$

and Δ_c given by (7.13).

Evaluation of the exponential factor of (7.15) is not difficult, especially in view of the algebraic simplifications of form

$$\sum_\mu \gamma_\mu(\gamma\cdot a)\gamma_\mu = 0,$$

peculiar to two dimensions. One finds

$$S_c'(x) = \exp[-\frac{i}{4\pi}\int d^2Q\left(\frac{1}{Q^2-i\varepsilon} - \frac{1}{Q^2+e^2/\pi-i\varepsilon}\right)\cdot$$

$$\cdot(1-e^{iQ\cdot x})]S_c(x), \tag{7.16}$$

a result which is completely finite, having no infrared or ultraviolet divergences. The integrals of (7.16) may be evaluated in terms of elementary functions; for $x^2 > 0$, one obtains

$$\ell n[S_c' \cdot S_c^{-1}] = \frac{1}{2} [K_o\left(\frac{xe}{\sqrt{\pi}}\right) + \gamma + \ell n\left(\frac{xe}{\sqrt{\pi}}\right)], \tag{7.17}$$

where γ is Euler's constant. As $x \to \infty$, the RHS of (7.17) increases logarithmically,

$$S_c'(x) \sim \left(\frac{xe}{\sqrt{\pi}}\right)^{1/2} \cdot e^{\gamma} \cdot S_c(x), \tag{7.18}$$

a behavior quite incompatible with that expected from (5.37). This difficulty is a consequence of the absence of bare mass; a statement equivalent to (7.18) is that $\tilde{S}_c'(p)$ does not have the pole structure $\sim (p^2)^{-1}$ near $p^2 = 0$ as does $\tilde{S}_c(p)$, but rather a cut of form $(p^2)^{-1-1/4}$. It is then not possible to define an S-matrix in terms of asymptotic fields, for there is no gap (such as the $\delta > 0$ of (5.19)) between the pole and onset of the continuum. Because the form of $S_c'(x-y)$ resembles that of (5.35), it is possible to define a gauge transformation of the third kind to a new gauge in which the exponential factor of (7.16) is exactly removed, and one clearly has a free-field result. Thus, if an S-matrix could be defined, it would be unity.

B. The Thirring Model

An economical way of reproducing all existing solutions to the Thirring model follows from the previous analysis. The Thirring Lagrangian is, in two-dimensional space-time,[3]

$$L' = -\frac{1}{2} g^2 \sum_{\mu} (\bar{\psi}\gamma_\mu\psi)(\bar{\psi}\gamma_\mu\psi), \tag{7.19}$$

with g a real, dimensionless coupling constant. The generating functional of the problem,

$$Z\{\eta,\bar{\eta}\} = \langle\left(\exp\,[i\int(\bar{\eta}\psi+\bar{\psi}\eta)]\right)_+\rangle,$$

has the formal solution

$$Z\{\eta,\bar{\eta}\} = N^{-1}\cdot\exp[-\frac{ig^2}{2}\int\left(\frac{\delta}{\delta\eta}\gamma_\mu\frac{\delta}{\delta\bar{\eta}}\right)\left(\frac{\delta}{\delta\eta}\gamma_\mu\frac{\delta}{\delta\bar{\eta}}\right)]\cdot$$

$$\cdot\exp[i\int\bar{\eta}\,S_c\eta], \tag{7.20}$$

and may be rewritten with the aid of Birula's trick[4] as

$$Z = N^{-1}\cdot\exp\left[-i\int\frac{\delta}{\delta\eta}\left(-g\frac{\delta}{\delta j_\mu}\right)\gamma_\mu\frac{\delta}{\delta\bar{\eta}}\right]\cdot\exp[\frac{i}{2}\int j_\nu j_\nu]\cdot$$

$$\cdot\exp[i\int\bar{\eta}\,S_c\eta]\Big|_{j=0}, \tag{7.21}$$

since translation of the ficticious source just serves to reproduce (7.20). However, one may first operate upon the $\eta,\bar{\eta}$ dependence of (7.21), obtaining

$$Z = N^{-1}\exp\{i\int\bar{\eta}G[\frac{1}{i}\frac{\delta}{\delta j}]\eta + L[\frac{1}{i}\frac{\delta}{\delta j}]\}\cdot\exp[\frac{i}{2}\int j_\nu j_\nu]\Big|_{j=0}$$

$$= N^{-1}\exp[-\frac{i}{2}\int\frac{\delta}{\delta A_\nu}\frac{\delta}{\delta A_\nu}]\cdot\exp\{i\int\bar{\eta}G[A]\eta+L[A]\}\Big|_{A=0}, \tag{7.22}$$

which closely resembles the QED generating functional, except that (i) the latter's photon propagator $D_c^{\mu\nu}(x-y)$ has been replaced by $\delta_{\mu\nu}\delta^{(2)}(x-y)$, and (ii) one has in (7.22) the instruction to let all photon sources vanish.

If the current operator, $\bar{\psi}\gamma_\mu\psi$, is properly defined, so that current conservation is maintained, there can be no physical objection to including, in (7.21), a coupling of the current to arbitrary longitudinal source dependence. This corresponds to a gauge transformation of the third kind, replacing the factor

$$\exp[\frac{i}{2}\int j_\mu j_\mu]$$

of (7.21) by

$$\exp[\frac{i}{2}\int j_\mu \Pi^{(\zeta)}_{\mu\nu} j_\nu,$$

where

$$\Pi^{(\zeta)}_{\mu\nu} = \delta_{\mu\nu}-\zeta(k^2)(k_\mu k_\nu/k^2-i\varepsilon)$$

in momentum space or

$$\delta_{\mu\nu}-\zeta(-\partial^2)(\partial_\mu\partial_\nu/\partial^2+i\varepsilon)$$

in configuration space. For the moment, ζ will be considered as a constant; for $\zeta = 1$, $\Pi^{(1)}_{\mu\nu} = \Pi_{\mu\nu}$, the familiar projection operator.

Equation (7.22) is then replaced by

$$Z = N^{-1} \exp[-\frac{i}{2}\int \frac{\delta}{\delta A_\mu} \Pi^{(\zeta)}_{\mu\nu} \frac{\delta}{\delta A_\nu}] \cdot$$

$$\cdot \exp\{i\int\bar{\eta}G[A]\eta + L[A]\}\Big|_{A\to 0}, \tag{7.23}$$

and it is clear that all functional operations may be carried through for any n-point function, in a manner analogous to, and using the G[A] and L[A] of the previous section. For example, one easily obtains the relation corresponding to (7.16),

$$\ln[S'_c\cdot S_c^{-1}] = ig^2[\zeta-\frac{g^2/\pi}{1+g^2/\pi}]\cdot[D_c(0)-D_c(x-y)], \tag{7.24}$$

and may assign different numerical values to ζ in order to reproduce solutions previously found and described in the literature.[5] In particular, there exists a gauge defined by

$$\zeta = \zeta_y = \frac{g^2}{\pi}\left[1 + \frac{g^2}{\pi}\right]^{-1},$$

analogous to the Yennie gauge ($\zeta_y = -2$) of four-dimensional infrared electrodynamics, for which the entire interaction is removed. Finally, one may generalize to the case $\zeta = \zeta(k^2)$, for which (7.24) is replaced by

$$\ell n[S_c' \cdot S_c^{-1}] = \frac{ig^2}{(2\pi)^2} \int \frac{d^2k}{k^2 - i\varepsilon} [\zeta(k^2) - \zeta_y][1 - e^{ik\cdot(x-y)}]. \tag{7.25}$$

The choice $\zeta(k^2) = \zeta_y + \alpha m^2(k^2 + m^2 - i\varepsilon)^{-1}$, with α a dimensionless constant and m an arbitrary mass parameter, reproduces the QED solution (7.16). Both the Thirring model and two-dimensional electrodynamics provide examples of a gauge-dependent theory, with Green's functions quite similar to those found in the soft photon limit of four-dimensional QED.

C. Neopolitan Models

Caianiello and co-workers have introduced[6] a class of nonlocal models obtained by replacing the free-field configuration space boson propagator $\Delta_c(x-y)$, which enters into the defining Green's function equations of various theories, by the constant $i\kappa^2$, where κ^2 is a real positive number. Although lacking in physical content, this type of model does display the algebraic structure which might be expected in realistic theories, and has been used[7] (for $L' = -g^2A^4$) to exhibit n-point functions holomorphic in g^{-1}. The method is illustrated here in the simplest context, that of a boson-fermion interaction of form $L' = -g\bar{\psi}\psi A$. The dressed fermion propagator, given by

$$S_c'(x-y) = \exp[-\frac{i}{2}\int \frac{\delta}{\delta A}\Delta_c\frac{\delta}{\delta A}]\cdot G(x,y|A)\cdot N^{-1}\exp L[A]\Big|_{A\to 0},$$

may be put into a simple parametric form when the replacement $\Delta_c(\kappa-\theta) \to i\kappa^2$ is made, for the functional differentiation operator becomes

$$\exp[\frac{\kappa^2}{2}\left(\int d^4u\,\frac{\delta}{\delta A(u)}\right)^2]$$

$$= \frac{1}{\sqrt{2\pi}}\int_{-\infty}^{+\infty} d\alpha\,\exp[-\frac{1}{2}\alpha^2+\alpha\kappa\int d^4u\,\frac{\delta}{\delta A(u)}],$$

and hence

$$S_c'(x-y) \Rightarrow \frac{1}{\sqrt{2\pi}}\int_{-\infty}^{+\infty} d\alpha\,\exp[-\alpha^2/2]G(x,y|\alpha\kappa)\cdot N^{-1}\exp L[\alpha\kappa], \tag{7.26}$$

where

$$G(x,y|\alpha\kappa) = \langle x|[m+\gamma\cdot\partial+\alpha\kappa g]^{-1}|y\rangle = S_c(x-y;m+\alpha g\kappa).$$

The closed-loop functional, in the absence of any specific redefinition, or removal of tadpole structure, will require a cut off both in configuration space ($\int d^4x \to VT$) and momentum space ($S_c(0;M)$ is quadratically divergent),

$$L[\alpha\kappa] \to \alpha\kappa(VT)\int_0^g dg'\,\mathrm{tr}[S_c(0;m+\alpha g'\kappa)].$$

However, since $L \sim i\alpha^2$ as $\alpha \to \infty$, there is no loss of generality if L is most simply omitted, and we now set $L = 0$, $N = 1$. In momentum space, (7.26) then takes on the compact form

$$\tilde{S}'_c(p) = \frac{1}{\sqrt{2\pi}} \int_{-\infty}^{+\infty} d\alpha \, \exp[-\tfrac{1}{2}\alpha^2] \, [m+\alpha\kappa g+i\gamma\cdot p]^{-1}, \qquad (7.27)$$

a representation which can be converted to one defining the error function Φ. In this example, the limit $g \to 0$ exists, with (7.27) reducing to the free propagator, $\tilde{S}_c(p)$.

D. The Lee Model

This partially soluble model has been extensively discussed[8] as a theory of model fields, the latter nonrelativistic and nonlocal. An alternate definition[9] may be obtained by performing simple approximations to the propagator functions which enter into the coupled Green's function equations of an exact, local, relativistic, nontrivial theory. In this way one sees clearly the special combinatoric structures contained in and representative of the model; e.g., one will always obtain linear equations for scattering amplitudes in the $V+n\theta$ sector, with these amplitudes coupled to those of smaller n, in contrast to the true field theoretic situation wherein the restriction to smaller n does not apply.

A detailed exposition from this point of view, and with the notation of the preceding chapters, may be found in Ref. 9, and hence may be omitted here. However, one must emphasize the historical importance of the Lee model in providing one of the first examples, albeit approximate, of interacting and explicitly renormalizable fields.

Notes

1. J. Schwinger, Phys. Rev. 128, 2425 (1962).

2. K. Johnson, Nuovo Cimento 20, 773 (1961), and C. Sommerfield, Ann. of Phys. 26, 1 (1963).

3. W. Thirring, Ann. of Phys. 3, 91 (1958).

4. I. Bialynicki-Birula, J. Math. Phys. 3, 1094 (1962).

5. H. M. Fried, Nuc. Phys. 75, 691 (1966).

6. E. R. Caianiello and A. Campolattaro, Nuovo Cimento XX, 1734 (1961).

7. E. R. Caianiello, A. Campolattaro and M. Marinaro, Nuovo Cimento XXXVIII, 113 (1965).

8. T. D. Lee, Phys. Rev. 95, 1392 (1954). An interesting variant of the usual model-field formulation has been given by M. Maxon and R. Curtis, Phys. Rev. 137, B996 (1965), using the LSZ formalism.

9. H. M. Fried, J. Math. Phys. 7, 583 (1966).

Chapter 8

NO-RECOIL METHODS

Those physically important situations in which a particle's mass--or energy--is very large compared to other relevant energies, may be treated by a generalization of the method briefly described in Chapter 5, Section A. The genesis of all such techniques lies in the Bloch-Nordsieck approximation of 1937, invented to provide a physical resolution of the so-called infrared catastrophe.[1] In its more modern contexts, Bloch-Nordsieck (B-N) methods may be applied, with profit, to an array of problems in high energy physics.

A. The Bloch-Nordsieck Approximation

We begin by treating the most useful case of fermion-vector boson interaction, $L' = ig\bar{\psi}\gamma\cdot A\psi$, although the bosons need not be massless, and the transition to scalar bosons may be made when desired. All the forms previously written for $G[A]$, $L[A]$, Z in QED are then valid here, and one first considers finding a representation for $G[A]$ in the limit of large fermion mass (or energy). Corresponding to this interaction, a fermion emits real and virtual quanta of the boson field to which it is coupled; and since 4-momentum conservation must hold in every such emission or absorption, the motion of each fermion must reflect this conservation. On the other hand, if the fermion mass, or energy, is much larger than the 4-momenta of each boson quanta, the fermion hardly suffers any recoil in every such event. This is precisely the infrared (IR) limit of QED, where only vanishingly small photon 4-momenta are considered. As explained in the appropriate sections, it is also a useful limit to apply in other cases, where fermion energy is large, and boson mass need not be zero. For all such cases, the essential physical approximation is the lack of fermion recoil.

The mathematical expression of this approximation is the replacement of the Dirac γ_μ matrices by c-number constants $-iv_\mu$, representing an averaged fermion 4-momentum, one unchanged by multiple soft emission and/or absorption. In accordance with the metric used throughout these pages, we associate v_μ of a BN Green's function $G_{BN}^{(v)}[A]$ with $(1/m)p_\mu$, where p_μ denotes the 4-momentum of the fermion so described, and $v^2 = -1$; in the fermion's rest frame, $v_o = +1$. The basic differential equation satisfied by $G[A]$ is now changed to read

$$\{m-iv\cdot[\partial^x-igA(x)]\}G_{BN}^{(v)}(x,y|A) = \delta^{(4)}(x-y), \tag{8.1}$$

and may be solved exactly for arbitrary $A(x)$. One proceeds by the "method of the proper-time parameter," writing the parametric representation for G as[2] $(m \to m-i\varepsilon)$

$$G_{BN}^{(v)}(x,y|A) = i\int_0^\infty d\xi \exp\left(-i\xi\{m-iv\cdot[\partial^x-igA(x)]\}\right)\cdot\delta(x-y)$$

$$= i\int_0^\infty d\xi\ e^{-i\xi m}\cdot e^{-\xi v\cdot\partial^x}\ F(\xi;x)\cdot\delta(x-y), \tag{8.2}$$

where

$$F(\xi;x) \equiv e^{\xi v\cdot\partial}\cdot e^{-\xi v\cdot(\partial-igA)}. \tag{8.3}$$

Had the replacements $\gamma_\mu \to -iv_\mu$ not been made, the F of (8.3) would depend upon the gradient operator ∂_x, as well as on x. That this is not the case here may be seen by calculating the differential relation obeyed by F,

$$\frac{\partial F}{\partial\xi} = ig\ e^{\xi v\cdot\partial}v\cdot A\ e^{-\xi v\cdot(\partial-igA)} = ig\ e^{\xi v\cdot\partial}v\cdot A\ e^{-\xi v\cdot\partial}F \tag{8.4}$$

or

$$\frac{\partial F(\xi;x)}{\partial \xi} = igv\cdot A(x+\xi v)\cdot F(\xi;x), \tag{8.5}$$

where (8.5) follows from (8.4) because $e^{\pm\xi v\cdot\partial}$ is a simple translation operator. With the boundary condition $F(\xi=0) = 1$, obvious from (8.3), one may write the solution

$$F(\xi;x) = \exp[ig\int_0^{\xi} d\xi' v\cdot A(x+\xi' v)],$$

and hence

$$G_{BN}^{(v)}(x,y|A) = i\int_0^{\infty} d\xi\ e^{-i\xi m}\delta(x-y-\xi v)\ \cdot$$

$$\cdot\ \exp[ig\int_0^{\xi} d\xi_1 v\cdot A(x-\xi_1 v)]. \tag{8.6}$$

Note that (8.6) reduces to the retarded function of (5.3) in the fermion's rest frame. In any Lorentz frame it is a retarded function, and hence the $L[A]$ constructed from (8.6) must vanish.

Had the replacements $\gamma_\mu \to -iv_\mu$ not been made, (8.4) would take the form

$$\frac{\partial F}{\partial \xi} = -g\left(e^{i\xi\gamma\cdot\partial}\gamma\cdot A\ e^{-i\xi\gamma\cdot\partial}\right)F, \tag{8.7}$$

leading to the exact representation for $G[A]$,

$$G[x,y|A] = i\int_0^{\infty} d\xi\ e^{-i\xi(m+\gamma\cdot\partial)}\ \cdot$$

$$\cdot\left(\exp[ig\int_0^{\xi} d\xi' e^{i\xi'\gamma\cdot\partial}(i\gamma\cdot A)e^{-\xi'\gamma\cdot\partial}]\right)_+\cdot\delta(x-y), \tag{8.8}$$

where $(\ )_+$ denotes a "time-ordering" of the variables ξ'. While exact, (8.8) is far too complicated to be useful in a nonperturbative sense. Another exact representation for $G[A]$, obtained by writing $G = [m-\gamma\cdot(\partial-igA)]G[A]$ and studying a sequence of no-recoil approximations for $G[A]$, has been constructed by Fradkin,[3] with leading terms resembling (8.6) but somewhat closer to those obtained in a boson model. The latter may be obtained (e.g., for the boson-boson interaction $L' = -(g^2/2)\psi^2 A$) by approximating solutions to the boson Green's function equation,

$$[m^2-\partial^2+gA]G[A] = 1, \tag{8.9}$$

where (8.9) is written in a formal, coordinate-independent way. Again, with $m \to m-i\varepsilon$, a representation may be written in the form

$$G(x,y|A) = \langle x|G[A]|y\rangle = (2\pi)^{-4}\int d^4p\ e^{ip\cdot x}\ \langle p|G|y\rangle,$$

where

$$\langle p|G|y\rangle = i\int_0^\infty d\xi\ e^{-i\xi(m^2+p^2)}\langle p|F(\xi)|y\rangle, \tag{8.10}$$

and $F(\xi) = e^{-i\xi\partial^2}\cdot e^{+i\xi[\partial^2-gA]}$. Here, as in Chapter 3, Section D, $\langle x|A|y\rangle = \delta(x-y)A(x)$, $\langle p|A|p+k\rangle = \tilde{A}(-k)$, and $\langle p|f(-\partial^2) = \langle p|f(p^2)$. A differential relation

$$\langle p|\frac{\partial F}{\partial \xi}|y\rangle = -ig\langle p|e^{-i\xi\partial^2}A\ e^{i\xi\partial^2}F|y\rangle$$

$$= -ig\int d^4k\ \tilde{A}(-k)e^{-i\xi(k^2+2k\cdot p)}\langle p+k|F(\xi)|y\rangle \tag{8.11}$$

may be written, from which it is convenient to extract the $\xi = 0$ phase dependence, setting $\langle p|F|y\rangle = e^{-ip\cdot y}f(\xi;p;y)$, so that

$$\frac{\partial f}{\partial \xi}(\xi;p;y) = -ig\int d^4k\tilde{A}(-k)e^{-ik\cdot y}\cdot e^{-i\xi(k^2+2k\cdot p)}f(\xi;p+k;y), \tag{8.12}$$

with $f(\xi=0) = 1$. The explicit y dependence on the RHS of (8.12) guarantees 4-momentum conservation in any process involving bosons described by $\tilde{A}(-k)$. Using the Baker-Hausdorf relation (Appendix D), an alternate way of writing (8.12) is

$$\frac{\partial f}{\partial \xi} = -ig\int dk\ \tilde{A}(-k)e^{-ik\cdot y}\cdot \exp[-2i\xi k\cdot p+k\cdot\frac{\partial}{\partial p}]\cdot f(\xi;p;y)$$

$$= -igA\left(y+2\xi p-i\ \frac{\partial}{\partial p}\right)\ f(\xi;p;y). \tag{8.13}$$

A sequence of BN approximations may now be defined by assuming small recoil of the ψ-boson, with the initial step the neglect of the $-i(\partial/\partial p)$ operator of (8.13); the result is an approximate potential-theory Green's function

$$G_{BN}(x,y|A) = i(2\pi)^{-4}\int d^4p e^{ip\cdot(x-y)}\int_0^\infty d\xi e^{-i\xi(m^2+p^2)}\cdot$$

$$\cdot\ \exp[-ig\int_0^\xi d\xi' A(y+2\xi' p)], \tag{8.14}$$

which should be compared with the fermion form (8.6). An alternate sequence of approximations may be defined directly from (8.12), with the leading term obtained by neglecting the k-dependence inside $f(\xi;p+k;y)$, while retaining the k^2 part of the multiplicative exponential factor; one then finds

$$G_{BN}(x,y|A) = i\int \frac{d^4p}{(2\pi)^4}\ e^{ip\cdot(x-y)}\int_0^\infty d\xi e^{-i\xi(m^2+p^2)}\cdot$$

$$\cdot \exp[-ig\int_0^{\xi} d\xi' \int \frac{d^4k}{(2\pi)^4} \tilde{A}(-k)e^{-ik\cdot y} \cdot$$

$$\cdot e^{-i\xi'(k^2+2k\cdot p)}]. \qquad (8.15)$$

Equations (8.14) and (8.15) are equivalent as long as one is dealing with soft boson momentum k_μ such that $|k^2| << |k\cdot p|$ for any process. For larger k, it turns out that the renormalized integrals constructed from (8.15) will automatically cut off at $k \sim m$ or $k \sim |t|^{1/2}$ whichever is largest, where $(-t)$ denotes some appropriate kinematical invariant of the particular problem under consideration; the same forms constructed from (8.6) or (8.14) will require an upper k cut off (logarithmic) in virtual processes, which must then be supplied by hand. For purposes of obtaining qualitative, few-parameter fits to high energy data of various sorts, it is easier to work with the simpler BN approximations of (8.6) or (8.14).

These forms find application in different high energy processes because it is there possible to produce large numbers of massive, but low energy bosons, in rough analogy to the soft photon possibilities of QED. One may consider, e.g., Feynman graphs with nucleon legs emitting many low-energy vector mesons, which through unitarity, can act to damp out related elastic amplitudes. In addition to such bremsstrahlung effects, there are the so-called multiperipheral processes, where mesons are emitted from the innards of a graph, rather than from its legs. In any functional analysis, a preliminary and most useful statement may be made in those situations where self-energy corrections to the particle emitting soft bosons may be neglected. In principle, of course, one must calculate all radiative corrections; but, as is demonstrated below, those corresponding to self-energy structure of external legs serve only to shift the bare mass to its physical value, while multiplying the entire Green's function by an appropriate renormalization constant. From the exact representation (8.10), with

$$\langle p|\mathcal{F}(\xi)|y\rangle \equiv e^{-ip\cdot y} f(\xi;p;y),$$

one obtains

$$K_x \mathcal{G}(x,y|A) = -(2\pi)^{-4}\int d^4q e^{iq\cdot(x-y)}\int_0^\infty d\xi f(\xi;q;y)\cdot$$

$$\cdot\ [\frac{\partial}{\partial\xi}\, e^{-i\xi(m^2+q^2)}]$$

$$= \delta(x-y)+(2\pi)^{-4}\int d^4q e^{iq\cdot(x-y)}\cdot$$

$$\cdot\int_0^\infty d\xi e^{-i\xi(m^2+q^2)}\,\frac{\partial}{\partial\xi}\, f(\xi;q;y), \qquad (8.16)$$

after an obvious integration-by-parts. Treating m as the physical mass, on the mass shell, $m^2+p^2 = 0$, there results

$$\int d^4x\ e^{-ip\cdot x} K_x \mathcal{G}(x,y|A)\Big|_{M.Sh.} = e^{-ip\cdot y} f(\infty;p;y), \qquad (8.17)$$

which is the desired relation. Similar computations may be carried through for the exact fermion Green's function; in the context of the BN model (8.6), these mass shell amputated forms may be written as

$$\int d^4x\ e^{ip\cdot x} G_{BN}^{(v)}(u,x|A)(m+v\cdot p)\Big|_{M.Sh.}$$

$$= e^{ip\cdot u}\ \exp[ig\int_0^\infty d\xi v\cdot A(u-\xi v)], \qquad (8.18)$$

and

$$\int d^4y\ e^{-ip'\cdot y}(m+v'\cdot p')G_{BN}^{(v')}(y,w|A)\Big|_{M.Sh.}$$

$$= e^{-ip'\cdot w}\ \exp[ig\int_0^\infty d\xi v'\cdot A(w+\xi v')], \qquad (8.19)$$

expressions which subsequently will be most useful.

B. Soft Photons: Fermion Self-Energy Structure

The problem of this section is the detailed computation of the self-energy structure of a charged fermion due to its cloud of soft, virtual photons. Perhaps the most precise and systematic formulation is that of Fradkin, but in the interest of simplicity we shall work with the BN Green's function (8.6), in terms of which $L[A] = 0$, $N = 1$, and

$$S'_{c,BN}(x-y) = i(2\pi)^{-4}\int d^4p\; e^{ip\cdot(x-y)}\int_0^\infty d\xi e^{-i\xi(m+v\cdot p)}\cdot$$

$$\cdot \exp[-\frac{i}{2}\int\frac{\delta}{\delta A_\mu}\, D^{(\zeta)}_{c,\mu\nu}\,\frac{\delta}{\delta A_\nu}]\exp[i\int f_\sigma A_\sigma]\Big|_{A=0}, \tag{8.20}$$

where

$$f_\sigma(z) = e\int_0^\xi d\xi_1 v_\sigma \delta(z-x+\xi_1 v),$$

and

$$\tilde{D}^{(\zeta)}_{c,\mu\nu}(k) = \left(\delta_{\mu\nu} - \zeta\,\frac{k_\mu k_\nu}{k^2-i\varepsilon}\right)\tilde{D}_c(k),$$

with the parameter ζ expressing the arbitrariness of gauge. The functional operations of (8.20) may, of course, be carried through, and one finds

$$\tilde{S}'_{c,BN}(p) = i\int_0^\infty d\xi e^{-i\xi[m+v\cdot p]}\cdot e^{\psi(\xi)} \tag{8.21}$$

with

$$\psi(\xi) = i\int f_\mu D^{(\zeta)}_{c,\mu\nu} f_\nu$$

$$= -\frac{e^2}{4\pi^2}\left[1+\frac{\zeta}{2}\right]\int_0^\xi d\xi_1 \int_0^{\xi_1} d\xi_2 [(\xi_1-\xi_2)^2 - i\varepsilon]^{-1}. \quad (8.22)$$

Passage to the form of $\psi(\xi)$ shown in (8.22) is dependent upon the restriction $v^2 = -1$. One immediately sees that this result is (properly) gauge dependent, and in the extreme sense: there exists a gauge, the Yennie gauge[4] with $\zeta = -2$, in which the entire effect is removed.

The integrals exhibited in (8.22) contain an unavoidable divergence when $\xi_1 = \xi_2$, a reflection of those ultraviolet divergences which would be met if the $\xi_{1,2}$ integrals were performed before $\int d^4k$, rather than in the present sequence.[5] However, regularization may be performed in configuration space as well as (and in this case more conveniently than) in momentum space; here, this is most simply accomplished by replacing $(\xi_1-\xi_2)^2-i\varepsilon$ of (8.22) by $(\xi_1-\xi_2-i\varepsilon)^2$, valid since $\xi_1 \geq \xi_2$, together with the identification $\varepsilon \sim \Lambda^{-1}$, where Λ corresponds to a virtual momentum cut off. The integrals of (8.22) are then elementary, and yield

$$\psi(\xi) \sim \frac{e^2}{4\pi^2}\left[1+\frac{\zeta}{2}\right]\{-i\xi\Lambda + \ln(\xi\Lambda)\}, \quad (8.23)$$

retaining only the leading, divergent, terms in $\xi\Lambda$. As in the computation of (5.10), that term of (8.23) linear in ξ may be interpreted as a renormalization of the bare mass appearing in (8.21), with

$$m_r \equiv m + \frac{e^2}{4\pi^2}\left[1+\frac{\zeta}{2}\right]\Lambda$$

denoting the new physical mass. It must be remarked that such a gauge-dependent mass renormalization occurs only because of the special, no-recoil approximations of the model. Its linear divergence results from the absence of negative energy states (a retarded function was used for S_c), and the consequent removal of a symmetry which should act to lower the degree of divergence to logarithmic. As a consequence of the Ward identity, it may be shown that realistic mass renormalization is independent of gauge. In all subsequent computations, each mass renormalization will be assumed to have been (correctly) performed, and the renormalized mass m_r simply written as m.

In the remainder of (8.23), we may write $\ln(\xi\Lambda) = \ln(\xi m) + \ln(\Lambda/m)$, identifying the constant logarithmic divergence as the log of the wave-function renormalization constant,

$$\ln Z_2 = \frac{e^2}{4\pi^2} [1+ \frac{\zeta}{2}] \ln(\Lambda/m), \tag{8.24}$$

a result which is (properly) gauge variant. The renormalized propagator

$$\tilde{S}'_{c,R}(p) \equiv Z_2^{-1} \tilde{S}'_c(p)$$

is then, with $\gamma = \frac{e^2}{4\pi^2} [1+ \frac{\zeta}{2}]$,

$$\tilde{S}'_{c,R}(p) = i\int_0^\infty d\xi (\xi m)^\gamma e^{-i\xi[m+v\cdot p]}$$

$$= (-im)^\gamma \Gamma(\gamma+1)[m+v\cdot p]^{-(1+\gamma)}, \tag{8.25}$$

illustrating the type of gauge dependent structure previously seen in two-dimensional QED. The mass shell in this model corresponds to $m+v\cdot p = 0$, and $\tilde{S}'_c(p)$ there has a simple pole in this variable only in the Yennie gauge. Hence, if one wishes to perform soft-photon computations in an S-matrix context without the use of a small photon mass, one should stay within the Yennie gauge.

For later use, it will be convenient to write the integral form of this model Z_2, which may be read off from (5.10) or (8.22),

$$\ln Z_2^{(\zeta=0)} = +\frac{4ie^2}{(2\pi)^4}\int \frac{d^4k}{k^2+\mu^2-i\varepsilon}\,\frac{p^2}{(2k\cdot p+i\varepsilon)^2}\,, \qquad (8.26)$$

where $p^2+m^2 = 0$, and a vector meson mass μ has been incorporated to avoid a logarithmic IR divergence (which, however, cannot be distinguished from the UV divergence, since for $\mu = 0$, $\ln Z_2 \sim \int_0^\infty dk/k$). It may be noted that the entire $\psi(\xi)$ of (8.22) has no IR singularity, with the latter appearing only when ψ is separated into a renormalization term plus remainder.

C. Soft Photons: Cancellation of IR Divergences

We now discuss the canonical IR problem, the removal of all IR divergences from a properly defined cross section. This topic is particularly suited to functional methods, since all operations may be carried through exactly for an arbitrary process. Indeed, a variant due to Schwinger[6] has been developed by Mahanthappa,[7] wherein one employs functional methods directly to formulate probabilities, rather than first calculating amplitudes; and the IR catastrophes never appear. Most of the practical work on this problem has been due to Yennie and co-workers,[8] and to Tsai,[9] who have specialized in extracting the soft electromagnetic radiative corrections necessary for a proper understanding of nonelectromagnetic interactions of various high energy scattering processes. Their final, finite forms may be obtained by the procedure outlined below, if the BN approximation of (8.6) is replaced by Fradkin's corresponding initial approximation; for scattering of charged particles with large momentum transfer, the two methods produce slightly different answers, corresponding to different ways of including photon momenta larger than infrared. Both methods agree in the nonrelativistic limit, however, and in both formulations the IR divergences are removed in the same way.

Complete cancellation of all IR divergences will be illustrated here in the simplest context of the scattering of a charged particle by a weak external field, but to all orders in its coupling to the radiation field. From (2.29), the amplitude for scattering with the production of photons of momenta $k_1 \cdots k_n$ is easily seen to be

$$<p',k_1 \cdots k_n|S|p>$$

$$= iZ_2^{-1} \cdot Z_3^{-n/2} (2\pi)^{-3/2(n+2)} \frac{1}{\sqrt{n!}} \varepsilon_{\mu_1}(k_1) \cdots \varepsilon_{\mu_n}(k_n) [\frac{m^2}{EE'}]^{1/2} \cdot$$

$$\cdot [2\omega_1 \cdots 2\omega_n]^{-1/2} \cdot \int d^4x' e^{-ip'\cdot x'} \int d^4x e^{ip\cdot x} \cdot$$

$$\cdot \int d^4z_1 e^{-ik_1\cdot z_1} \cdots \int d^4z_n e^{-ik_n\cdot z_n} \frac{\delta}{\delta A_{\mu_1}(z_1)} \cdots \frac{\delta}{\delta A_{\mu_n}(z_n)} \cdot$$

$$\cdot \exp[-\frac{i}{2} \int \frac{\delta}{\delta A} D_c \frac{\delta}{\delta A}] \cdot$$

$$\cdot \left(\bar{u}(p')\vec{\mathcal{D}}_{x'}, G(x',x|A+A^{ext})\overleftarrow{\bar{\mathcal{D}}}_x u(p)\right)\Big|_{A\to 0}, \tag{8.27}$$

and all closed fermion loops have been dropped in anticipation of the BN approximation, (8.6), to be made for the electron propagators. To first order in A^{ext},

$$G(x',x|A+A^{ext}) \simeq G(x',x|A) + ie\int G(x',u|A)\gamma\cdot A^{ext}(u)G(u,x|A), \tag{8.28}$$

where the first term on the RHS of (8.28) vanishes because of 4-momentum conservation, $p' \neq p+k_1+ \cdots +k_n$.

We now specialize to the case of soft photons, possessing no induced structure of their own ($L = 0$, $Z_3 = 1$), writing

$$ie\int G_{BN}^{(v')}(x',u|A)\gamma\cdot A^{ext}(u)G_{BN}^{(v)}(u,x|A)$$

for the quantities of (8.28), where $v = p/m$ and $v' = p'/m$; this corresponds to a (classical) electron scattering from x to x', with a corresponding momentum change from p to p'. The last line of (8.27) then becomes

$$\exp[-\frac{i}{2}\int \frac{\delta}{\delta A} D_c \frac{\delta}{\delta A}](ie)\left(\bar{u}(p')\gamma_\nu u(p)\right)\int d^4u A_\nu^{ext}(u)\cdot$$

$$\cdot e^{i(p-p')\cdot u}\cdot\exp[i\int f\cdot A], \qquad (8.29)$$

where

$$f_\mu(z) = e\{v_\mu\int_0^\infty d\xi\delta(z-u+\xi v) + v'_\mu\int_0^\infty d\eta\delta(z-u-\eta v')\},$$

and use has been made of (8.18) and (8.19). Calculating in the Feynman gauge, the functional operations yield the familiar form,

$$\exp[\frac{i}{2}\int fD_c f + i\int f\cdot A],$$

where

$$\exp[\frac{i}{2}\int fD_c f] = \frac{i}{2}e^2v^2\int_0^\infty d\xi_1\int_0^\infty d\xi_2 D_c([\xi_1-\xi_2]v) +$$

$$+\frac{i}{2}e^2v'^2\int_0^\infty d\eta_1\int_0^\infty d\eta_2 D_c([\eta_1-\eta_2]v') +$$

$$+ ie^2(v\cdot v')\int_0^\infty d\xi\int_0^\infty d\eta\ D_c(\xi v+\eta v'), \qquad (8.30)$$

and A is not allowed to vanish until those differentiations corresponding to real photon production are performed. The first and second terms on the RHS of (8.30) are associated with self-energy structure of the incoming and outgoing electron propagators. They each contain the same linear divergence corresponding to the fermion mass shifts, which have been ignored when using (8.18), (8.19); clearly, had mass renormalization been performed in its proper sequence, these divergences would be missing, and we henceforth omit all such dependence. Remaining in both terms is the same logarithmically divergent integral, exactly given by (8.26), and providing two factors of $\ell n\ Z_2$. Precisely at this stage the Z_2^{-1} factor, appearing in the original definition of the S-matrix element, enters and serves to remove one factor of $\ell n\ Z_2$, giving a complete exponent as the sum of the third RHS term of (8.30)--the cross-linkage dependence, containing all soft photon propagators linking the initial and final electron legs--plus a single $\ell n\ Z_2$,

$$- \frac{ie^2}{(2\pi)^4} \int \frac{d^4k}{k^2-i\varepsilon} \left\{\frac{p\cdot p'}{(k\cdot p'+i\varepsilon)(k\cdot p+i\varepsilon)} - \frac{p^2}{(k\cdot p+i\varepsilon)^2}\right\}$$

$$= \frac{ie^2}{2(2\pi)^4} \int \frac{d^4k}{k^2-i\varepsilon} \left(\frac{p}{k\cdot p} - \frac{p'}{k\cdot p'}\right)^2, \tag{8.31}$$

where $p^2 = p'^2 = -m^2$. Because the forms (8.6) rather than (8.15) have been employed, (8.31) has a logarithmic UV divergence, which is to be removed by limiting the magnitude of the virtual photon momenta. For small photon momenta, the exponent of (8.31) represents the renormalized vertex function, in its dependence on soft virtual photons to all orders in e^2.

Evaluation of (8.31) is simplest in the nonrelativistic limit, $|\vec{p}| << m >> |\vec{p}'|$, and gives $(q = p-p')$

$$- \frac{\alpha}{3\pi} \left(\frac{q^2}{m^2}\right) \int_{k_{min}}^{m} \frac{d\omega}{\omega}, \tag{8.32}$$

where $\alpha = e^2/4\pi$ and cut offs m and k_{min} have been inserted to limit the magnitude of the soft photons included. In the relativistic limit (e.g., Chapter 9, Section B), the factor of q^2/m^2 exhibited in (8.32) is changed to one proportional to $\ln(q^2/m^2)$, while an upper cut off somewhat larger than m may be expected; for large momentum transfer processes (made possible by A^{ext}, not by the multiple soft photon exchanges) the relevant momentum scale for soft effects can increase. This question is left open by the present discussion, which describes the IR limit but not the approach to that limit. It is answered in a definite if somewhat arbitrary way by the Fradkin-Yennie-Tsai treatments, in which (8.31) is replaced by

$$\frac{ie^2}{2(2\pi)^4}\int\frac{d^4k}{k^2}\left(\frac{2p+k}{k^2+2k\cdot p}-\frac{2p'+k}{k^2+2k\cdot p'}\right)^2.$$

No upper cut off is needed here, for this renormalized integral is, in the limit of large momentum transfer, cut off for $k \sim \sqrt{q^2}$, and hence produces a net exponential factor proportional to $-\ln^2(q^2)$. For nonzero photon mass μ, the lower limit k_{min} may (to within additive constants) be replaced by μ, and one obtains the behavior

$$\Gamma_\mu \sim \gamma_\mu \exp[-G \ln^2(q^2/\mu^2)]$$

recently noted by Jackiw[10] as the sum of the leading $\ln(q^2)$ dependence in every order of perturbation theory for the vertex function in massive photon QED. Quite apart from the details of the soft photons' upper cut off, the essential physical picture which emerges is one of strong damping of an elastic amplitude by the virtual exchange of soft vector mesons. By unitarity, one expects this elastic behavior to be related to the growth of inelastic amplitudes, a physical effect underlying the 1937 BN treatment of soft photons, and providing impetus for current, nonelectromagnetic studies at high energies.

The evaluation of $\langle p',k_1\cdots k_n|S|p\rangle$ may be completed by performing those functional differentiations corresponding to real soft photon production; here, each operation

$$\int d^4z \; e^{-k\cdot z} \; \varepsilon_\mu \frac{\delta}{\delta A_\mu(z)}$$

upon $\exp[i\int f\cdot A]$ simply yields a factor

$$e[\frac{\varepsilon\cdot p'}{k\cdot p'} - \frac{\varepsilon\cdot p}{k\cdot p}] \; e^{-ik\cdot u}.$$

The calculation of

$$\int d^4u \; A_\nu^{ext}(u) \; \exp\left(i[p-p'-\sum_\ell k_\ell]\cdot u\right)$$

then produces

$$(2\pi)^4 \; \tilde{A}_\nu^{ext}(p' + \sum_\ell k_\ell - p')$$

which we replace by

$$(2\pi)^4 \; \tilde{A}_\nu^{ext}(p'-p)$$

in this IR limit. In terms of the corresponding amplitude

$$\langle p'|S|p\rangle = -(2\pi)e[\frac{m^2}{EE'}]^{1/2}\left(\bar{u}(p')\gamma\cdot\tilde{A}^{ext}(p'-p)u(p)\right),$$

describing the scattering of an electron by a weak external field in the absence of the radiation field, (8.27) may be written in the nonrelativistic, soft photon limit as

$$\langle p',k_1\cdots k_n|S|p\rangle = \langle p'|S|p\rangle \cdot \exp[-\frac{\alpha}{3\pi}\left(\frac{q^2}{m^2}\right)\ell n\left(\frac{m}{k_{min}}\right)]\cdot$$

$$\cdot \frac{e^n}{\sqrt{n!}} (2\pi)^{-3n/2} \prod_{\ell=1}^{n} \frac{1}{\sqrt{2\omega_\ell}} \cdot$$

$$\cdot \left[\frac{\varepsilon_\ell \cdot p'}{k_\ell \cdot p'} - \frac{\varepsilon_\ell \cdot p}{k_\ell \cdot p}\right] . \qquad (8.33)$$

The probability for the emission of n soft photons is given by the absolute square of (8.33). However, in such a process one cannot experimentally detect photons of arbitrarily low energy, which means that one must be careful to ask the physically correct question: what is the probability for an electron to scatter from p to p' with the production of an arbitrary number of soft, unmeasurable photons. Denoting the energy resolution of the particular experiment by ΔE (no photon of energy less than ΔE can be detected), one obtains for the production probability of n soft photons as a result of the scattering,

$$P_n(\Delta E) = \exp[- \frac{2\alpha}{3\pi} \left(\frac{q^2}{m^2}\right) \ln\left(\frac{m}{k_{min}}\right)] |\langle p'|S|p\rangle|^2 \cdot$$

$$\cdot \frac{e^{2n}}{n!} \prod_{\ell=1}^{n} \sum_{\lambda_\ell=1}^{2} \int_0^{\Delta E} \frac{d^3k_\ell}{(2\pi)^3} \frac{1}{2\omega_\ell} \left[\frac{\varepsilon_\ell \cdot p'}{k_\ell \cdot p'} - \frac{\varepsilon_\ell \cdot p}{k_\ell \cdot p}\right]^2 , \qquad (8.35)$$

where $\sum_{\lambda_\ell}$ denotes a sum over both polarizations of the ℓth photon. The computations are again simplified in the nonrelativistic region, and one easily finds

$$\frac{e^2}{(2\pi)^3} \sum_{\lambda=1}^{2} \int_0^{\Delta E} \frac{d^3k}{2\omega} \left[\frac{\varepsilon \cdot p'}{k \cdot p'} - \frac{\varepsilon \cdot p}{k \cdot p'}\right]^2 = \frac{2\alpha}{3\pi}\left(\frac{q^2}{m^2}\right) \ln\left(\frac{\Delta E}{k_{min}}\right) , \qquad (8.35)$$

where k_{min} has again been used for each IR divergence. There follows the Poisson distribution in $\ell n\, k^{-1}_{min}$,

$$P_n(\Delta E) = \frac{1}{n!}\left[\frac{2\alpha}{3\pi}\left(\frac{q^2}{m^2}\right)\ell n\left(\frac{\Delta E}{k_{min}}\right)\right]^n \cdot$$

$$\cdot \exp\left[-\frac{2\alpha}{3\pi}\left(\frac{q^2}{m^2}\right)\ell n\left(\frac{m}{k_{min}}\right)\right]|\langle p'|S|p\rangle|^2, \qquad (8.36)$$

and one sees that the probability for the emission of a finite number of soft photons vanishes, as $k_{min} \to 0$. Summing over all n, however, produces the finite, physical quantity

$$\sum_{n=0}^{\infty} P_n(\Delta E) = |\langle p'|S|p\rangle|^2 \cdot \exp\left[-\frac{2\alpha}{3\pi}\left(\frac{q^2}{m^2}\right)\ell n\left(\frac{m}{\Delta E}\right)\right], \qquad (8.37)$$

which, for small q^2/m^2, is the result conjectured by Schwinger[11] and proved by Yennie and Suura.[12] In the relativistic limit, the Fradkin-Yennie-Tsai approach generates an answer in which the exponential of (8.37) is replaced by

$$-\frac{2\alpha}{\pi}\,\ell n\left(\frac{q^2}{m^2}\right)\ell n\left(\frac{E}{\Delta E}\right);$$

the details of these calculations are somewhat different than those of the simpler one carried out here, but the crucial cancellation of all IR divergences is the same.

D. Soft Pions

There are two broad categories which fall under this title. Of them, perhaps the most widely known[13] is that collection of theorems and approximations concerning low energy external pions, pions which appear among the external legs of a given Feynman graph. Results which have been obtained link the low energy pion limits of different reactions, and follow from assumptions of PCAC and current algebra, or equivalently from the degree of chirality of a given phenomenological Lagrangian (a Lagrangian used in tree-graph approximation only). The second category deals with the summation of many virtual pions, in the true sense of quantum fluctuations. While this subject is still in relative infancy, there has been some recent progress that is worth mentioning.

1) It has long been known that the static model of massive nucleon-nucleon interaction, by the exchange of <u>scalar</u> pions, is exactly soluble in the sense that corresponding scattering amplitudes containing all cross-linkages may be written down in closed parametric form. One simply develops an approximate $G_{BN}[\pi]$, given as the exponential of a linear form in $\pi(x)$, and is then able to perform all functional operations in the manner of Chapters 5, 8, and 9. The corresponding static nucleon Green's function with isotopics, $G_{BN}[\vec{\tau}\cdot\vec{\pi}] = [M-i\partial_0+g\vec{\tau}\cdot\vec{\pi}]^{-1}$, has never been obtained in closed form, because of the combinatoric complexity required to prevent charged pions from being emitted independently from the same nucleon line; e.g., after a proton emits a π^+, the next charged pion which the resultant neutron may emit must be a π^-. Recently, however, Fronsdal and Huff have obtained the exact potential energy function for the interaction of a pair of static nucleons, including isotopics.[14] Deviation from the Yukawa form, which is valid for large nucleon separation, becomes apparent at small distances. (As of this writing, the Fronsdal-Huff potential has not yet been applied to fit nucleon data.) Similar results have been obtained by them for the effective spin dependent potential between static nucleons in ps-ps theory.

2) Summation of the leading $\ell n(q^2)$ dependence, for large momentum transfer, in every order of perturbation theory of the ps-ps interaction $L' = ig\bar{\psi}\gamma_5\vec{\tau}\cdot\vec{\pi}\psi$ has been carried out for the nucleon electromagnetic form factors by Applequist and Primack.[15] Strictly speaking, this has little to do with soft pions, at least in the previous photon sense, because the P-wave nature of the pion-nucleon coupling suppresses the IR

region; there are no IR divergences in this theory as the pion mass is allowed to vanish. This calculation is of interest because of its unphysical results, in which expected damping does not occur.

Forms which specifically extract the "softest" dependence possible in the same ps theory, dependence following from low energy virtual, pion pair exchange, have been shown to produce form factor damping.[16] This calculation is perhaps of more interest from a functional point of view than because of its efficacy in fitting nucleon form factor data, for the combinatorics involved require the use of the full Eq. (3.59), rather than its simpler version with $B = 0$. Rough estimates of the damping resulting from this mechanism have been made, which alone compare rather poorly with experimental data. If the pions are made chiral, by embedding the calculation in the nonlinear σ model or by enforcing chirality in any other way, all the soft-pion-pair structure is exactly removed, to all orders in the strong coupling, g.

3) The absence of damping dependent upon appropriate kinematical invariants in chiral theories has been explicitly shown by Weinberg,[17] who produced an extraordinary solution of the complete soft-chiral-pion problem. As amended by Brown,[18] one finds an S-matrix with chiral-violating elements, built, e.g., out of chiral-violating nonsoft pions, that are damped by constant factors.

In summary, these calculations suggest that multiple soft-pion exchange plays no appreciable role in providing the strong kinematical damping seen in recent high-energy experiments. In subsequent sections, we shall concentrate instead on the simpler neutral-vector-meson (NVM) "gluon" models, which can serve to provide at least a qualitative description of different reactions.

Notes

1. F. Bloch and A. Nordsieck, Phys. Rev. 52, 54 (1937).

2. A discussion of this approximation is given in the text by Bogoluibov and Shirkov, Ref. 4, Chapter 1.

The derivation here follows that of Ref. 1, Chapter 2.

3. See Ref. 3, Chapter 3.

4. H. M. Fried and D. R. Yennie, Phys. Rev. 112, 1391 (1958).

5. A. V. Svidvinski, JETP 4, 179 (1957) (English translation).

6. J. Schwinger, J. Math. Phys. 2, 407 (1961).

7. K. Mahanthappa, Phys. Rev. 126, 329 (1962).

8. For example, D. R. Yennie, S. Frautschi, and H. Suura, Ann. Phys. (New York) 13, 379 (1961).

9. For example, L. W. Mo and Y. S. Tsai, Rev. Mod. Phys. 41, 205 (1969).

10. R. Jackiw, Ann. Phys. (New York) 48, 292 (1968).

11. J. Schwinger, Phys. Rev. 76, 790 (1949).

12. D. R. Yennie and H. Suura, Phys. Rev. 105, 1378 (1957).

13. An excellent review of this subject has been given by S. Weinberg, *Brandeis University Summer Lectures 1970*, Vol. 1, (see Ref. 3, Chapter 6).

14. C. Fronsdal and R. W. Huff, ICTP Preprints IC/71/119 and 120.

15. T. Applequist and J. R. Primack, Phys. Rev. D1, 1144 (1970).

16. H. M. Fried, Phys. Rev. D2, 3035 (1970); C. E. Carlson and T. L. Neff, Phys. Rev. D4, 532 (1971).

17. S. Weinberg, Phys. Rev. D2, 674 (1970); and 3085 (1970).

18. L. Brown, Phys. Rev. D2, 3083 (1970).

Chapter 9

RELATIVISTIC EIKONAL PHYSICS

This chapter is concerned primarily with the exploitation of a simple field-theoretic model of fermion-fermion scattering, wherein multiple NVM exchange takes place. It is not to be supposed that this gluon model is necessarily preferable to any other theory or mechanism of interactions at high energy (a variant of the A^3 theory is discussed in Chapter 10), nor is any fundamental significance attached to the particular NVM exchanged (which might be viewed as ρ_0 or ω). For certain technical, perhaps fortuitous reasons, this theory is particularly amenable to functional techniques, and yields results which compare favorably with much recent experimental data. In both this and the following chapter, many topics of high current interest have been omitted, or treated with what may seem the most callous indifference to involved and responsible authors. The reader should remember that these topics are still in a state of flux, and that the presentation here has been determined and arranged largely by considerations of relevance to the functional approach.

Following the techniques of Chapter 2, the S-matrix element for fermion-fermion elastic scattering ($p_1+p_2 \to p_1'+p_2'$) may be written in the form $S = 1+iT$, where

$$\langle p_1'p_2'|T|p_1p_2\rangle$$

$$= -iZ_2^{-2}(2\pi)^{-6}\left[\frac{m^2}{E_1E_2E_1'E_2'}\right]^{1/2}\int d^4x_1 e^{ip_1x_1}\int d^4x_2 e^{ip_2x_2}\cdot$$

$$\cdot\int d^4y_1 e^{-ip_1'y_1}\int d^4y_2 e^{-ip_2'y_2}\cdot\bar{u}_{S_1'}^{\alpha_1'}(p_1')\bar{u}_{S_2'}^{\alpha_2'}(p_2')\cdot u_{S_1}^{\beta_1}(p_1)u_{S_2}^{\beta_2}(p_2)\cdot$$

$$\cdot(\mathcal{D}_{y_1})_{\alpha_1'\beta_1'}(\mathcal{D}_{y_2})_{\alpha_2'\beta_2'}(\overline{\mathcal{D}}_{x_1})_{\alpha_1\beta_1}(\overline{\mathcal{D}}_{x_2})_{\alpha_2\beta_2}\cdot$$

$$\cdot \frac{\delta}{\delta\bar{\eta}_{\beta_1'}(y_1)} \cdot \frac{\delta}{\delta\bar{\eta}_{\beta_2'}(y_2)} \cdot \frac{\delta}{\delta\eta_{\alpha_1}(x_1)} \cdot \frac{\delta}{\delta\eta_{\alpha_2}(x_2)} \cdot (NZ) \Big|_{\substack{\eta=\bar{\eta}=0 \\ j=0}}, \tag{9.1}$$

and $Z\{j,\eta,\bar{\eta}\}$ denotes the familiar generating functional of QED, with the propagator of a massive NVM field replacing the photon propagator. Since current is again to be conserved, all the gauge properties of Chapter 5, Section C are valid here; in particular, we may work in the Feynman gauge, $\Delta_{c,\mu\nu} = \delta_{\mu\nu}\Delta_c$, even though the proper function of a free, massive, spin-one boson is $(\delta_{\mu\nu}-\partial_\mu\partial_\nu/\mu^2)\Delta_c$. Incidentally, this is a good example of the special cancellations and rearrangements of divergent quantities, which here change this nonrenormalizable theory (due to the $(\partial_\mu\partial_\nu/\mu^2)\Delta_c$ dependence) into a renormalizable theory (when only $\delta_{\mu\nu}\Delta_c$ is used), with all divergences associated with the nonrenormalizable behavior absorbed into the gauge-dependent $Z_{1,2}$.

Suppressing spinor indices, the last line of (9.1) may be written as

$$M(x_1y_1,x_2y_2) = i^2\exp[-\frac{i}{2}\int \frac{\delta}{\delta A}\Delta_c \frac{\delta}{\delta A}]\cdot G(y_1x_1|gA)\cdot$$

$$\cdot G(y_2,x_2|gA)\cdot \exp L[A]\Big|_{A\to 0}, \tag{9.2}$$

with $L[A] = \mathrm{Tr}\,\ell n[1-ig\gamma\cdot AS_c]$ and $A \equiv \int\Delta_c j$. More precisely, one should multiply (9.2) by a factor of 1/2, and add to it the contribution obtained by permuting the indices (y_1,β_1') and (y_2,β_2'); but for simplicity, we henceforth treat the nucleons as distinguishable (labeled by subscripts I, II) and omit the permutation.

A. Small-Angle Formalism

For that class of Feynman graphs without closed-fermion-loops, composed of multiple NVM exchange between a pair of fermions, it has been shown[1] that the large energy ($s = -(p_1+p_2)^2$), fixed momentum transfer ($t = -(p_1-p_1')^2$) limit, $s \to \infty$, $t/s \to 0$, is correctly described by treating the exchanged vector mesons as soft compared to the highly energetic fermions. More precisely, the leading $\ln s$ dependence in every order of perturbation theory (when all graphs of a given order in g^2 are summed, taking advantage of cancellations which occur between different graphs of the same order) may be reproduced by the perturbation expansion of the eikonal amplitude obtained below. This observation is an *a posteriori* argument for the validity of the initial, almost intuitive soft meson approximations, but one that appears to be restricted to NVM exchange only.[2]

To simplify the analysis, the eikonal derivation will here be given in the absence of closed loops ($L = 0$, $N = 1$), with the form of the latters' contribution appended at a later stage. Perhaps the simplest derivation of $T(g^2) \equiv \langle p_1'p_2'|T|p_1p_2\rangle$ follows from the calculation of $\partial T/\partial g^2$, for which one first forms the derivative with respect to coupling constant of (9.2),

$$\frac{\partial M}{\partial g} = -\exp[-\frac{i}{2}\int \frac{\delta}{\delta A}\Delta_c\frac{\delta}{\delta A}]\{\frac{\partial}{\partial g}G_I(y_1,x_1|gA)\cdot$$

$$\cdot G_{II}(y_2,x_2|gA) + G_I(y_1,x_1|gA)\frac{\partial}{\partial g}G_{II}(y_2,x_2|gA)\}. \tag{9.3}$$

The relation $\partial G/\partial g = i\int G\gamma\cdot AG$ may be used to rewrite each derivative of (9.3); both terms give an identical contribution, and so we may simply multiply the RHS of (9.3) by a factor of 2, and write

$$\frac{\partial M}{\partial g} = -2i\,\exp[-\frac{i}{2}\int \frac{\delta}{\delta A}\Delta_c\frac{\delta}{\delta A}]\cdot\int d^4z_1\big(G_I(y_1,z_1|gA)\gamma\cdot A(z_1)\cdot$$

$$\cdot G_I(z_1,x_1|gA)\big)\cdot G_{II}(y_2,x_2|gA)\Big|_{A\to 0}. \tag{9.4}$$

Passing the explicit $A_\mu(z_1)$ source through the differential operator, as in Chapter 6, Section C, the RHS of (9.4) yields,

$$(-2i)(-i)\exp[-\frac{i}{2}\int \frac{\delta}{\delta A}\Delta_c \frac{\delta}{\delta A}]\cdot\int d^4z_1\int d^4z_2\Delta_c(z_1-z_2)\cdot$$

$$\cdot\frac{\delta}{\delta A_\mu(z_2)}\cdot\{(G_I\gamma_\mu G_I)\cdot G_{II}\}\Big|_{A\to 0}$$

or

$$\frac{\partial M}{\partial g^2} = -i\int d^4z_1\int d^4z_2\Delta_c(z_1-z_2)\exp[-\frac{i}{2}\int\frac{\delta}{\delta A}\Delta_c\frac{\delta}{\delta A}]\cdot$$

$$\cdot\{(G_I\gamma_\mu G_I)(G_{II}\gamma_\mu G_{II})+2(G_I\gamma_\mu G_I\gamma_\mu G_I)\cdot G_{II}\}\Big|_{A\to 0}. \tag{9.5}$$

Except for the omission of closed fermion loops, (9.5) is an exact expression.

We now adopt the simplest no-recoil approximation of the previous chapter, replacing each G of (9.5) by a $G_{BN}^{(v)}$, $v_\mu \equiv p_\mu/m$, where p denotes the four-momentum of that fermion associated with the appropriate external coordinates; e.g.,

$$G_I(y_1,z_1|A) \to G_{I,BN}^{(p_1/m)}(y_1,z_1|A).$$

Mass shell amputation of (9.1) is performed according to (8.18), (8.19), while the remaining spinor factors may be evaluated in the limit $(p_1-p_1')/\sqrt{s}\to 0$ using $\bar{u}(p)\gamma_\mu u(p) = -ip_\mu/m$. With

$$\langle p_1'p_2'|T|p_1p_2\rangle = (2\pi)^{-2}\left[\frac{m^4}{E_1E_2E_1'E_2'}\right]^{1/2} M(g^2),$$

one then obtains

$$\frac{\partial M}{\partial g^2} = -\left(\frac{p_1 \cdot p_2}{m^2}\right)(2\pi)^{-4}\int d^4z_1 \int d^4z_2 \Delta_c(z_1-z_2) \cdot$$

$$\cdot \exp[i(p_1-p_1')\cdot z_1 + i(p_2-p_2')\cdot z_2] \cdot$$

$$\cdot Z_2^{-2} \cdot \exp[-\frac{i}{2}\int \frac{\delta}{\delta A}\Delta_c \frac{\delta}{\delta A}] \cdot \exp[i\int A[f_I+f_{II}]]\Big|_{A\to 0}, \tag{9.6}$$

where the single G_{II} term of (9.5) has been removed by the double, mass shell amputation, and

$$f^\mu_{I,II}(w) = g\int_0^\infty d\xi [p^\mu_{1,2}\delta(w-z_{1,2}+\xi p_{1,2}) + $$

$$+ p'^\mu_{1,2}\ \delta(w-z_{1,2}-\xi p'_{1,2})]. \tag{9.7}$$

The functional operation of (9.6) is elementary, and its results may be classified as self-energy terms plus soft NVM exchanges between different nucleon legs. Of the complete answer,

$$\exp[\frac{i}{2}\int (f_I+f_{II})\Delta_c(f_I+f_{II})],$$

each of the four factors of form

$$\exp[\frac{i}{2}\ g^2p^2\int_0^\infty\int d\xi d\eta\ \Delta_c([\xi-\eta]p)]$$

generates self-effects for the particular nucleon line, containing an unwanted mass renormalization plus a factor of $\ell n\ Z_2$. As in (8.30), each mass shift is to be dropped, while the four factors of $\ell n\ Z_2$ convert the Z_2^{-2} of (9.6) to a net factor of Z_2^{+2}. Of the cross-linkages, the terms

$$\exp[ig^2\int_0^\infty\int d\xi d\eta\{p_1\cdot p_1'\Delta_c(\xi p_1+\eta p_1')+p_2\cdot p_2'\Delta_c(\xi p_2+\eta p_2')\}]$$

represent just the product of the t-dependent vertex functions of this model, which are automatically renormalized by the net Z_2^{+2} dependence, in exactly the same manner as found in (8.31). For small momentum transfer, these renormalized combinations are small and may be neglected, and we shall follow this simplifying step in the remainder of this section. For moderate $-t \approx m^2$, this vertex dependence can be significant and should be retained;[3] for large $-t \gg m^2$, it is crucial and cannot be neglected.[4,5]

The remaining cross-linkage dependence has the form

$$\exp[i\int f_I^\mu\cdot\Delta_c\cdot f_{II}^\mu], \tag{9.8}$$

and corresponds to s- and u-channel soft NVM exchange. in the limit $t/s \to 0$, the $f_{I,II}$ of (9.7) may be replaced by

$$f_{I,II}^\mu(w) = g\int_{-\infty}^{+\infty} d\xi\ p_{1,2}^\mu\ \delta(w-z_{1,2}+\xi p_{1,2}), \tag{9.9}$$

and the exponent of (9.8) then simplifies to

$$ig^2(p_1\cdot p_2)\int_{-\infty}^{+\infty} d\xi\int_{-\infty}^{+\infty} d\eta\ \Delta_c(z_1-z_2-\xi p_1+\eta p_2). \tag{9.10}$$

Evaluation of (9.10) is simplest in the CM frame, with coordinate system chosen so that $\vec{p}_1 = -\vec{p}_2$ lies along the $\hat{z}$ direction. Inserting the standard Fourier representation for Δ_c, the integrals $\int d\xi\int d\eta$ may be performed, and yield, with $\gamma(s) = s-2m^2/\sqrt{s(s-4m^2)}$,

$$-i \frac{g^2}{(2\pi)^2} \gamma(s) \int \frac{d^2k}{k^2+\mu^2} e^{i\vec{k}\cdot\vec{z}} \tag{9.11}$$

where $\vec{z} = \vec{z}_1-\vec{z}_2$ is to take on the significance of a two-dimensional (z_x,z_y) impact parameter. It is important to note that the eikonal limits have provided dependence upon the variable $\vec{z}$ only, rather than upon the entire four-vector z_μ. Translational invariance may now be used to display energy-momentum conservation,

$$\frac{\partial}{\partial g^2} M_{eik} = -\left(\frac{p_1\cdot p_2}{m^2}\right)\delta^4(p_1+p_2-p_1'-p_2')\cdot\int d^4z\, e^{iq\cdot z}\, \Delta_c(z)\cdot$$

$$\cdot\exp[-i \frac{g^2}{(2\pi)^2} \gamma(s)\int \frac{d^2k}{k^2+\mu^2} e^{i\vec{k}\cdot\vec{z}}], \tag{9.12}$$

where $q = p_1-p_1'$. In the CM, $q_o = 0$, while $q_3 \to 0$ as $t/s \to 0$. Hence, upon inserting a representation for $\Delta_c(z)$, the $z_{0,3}$ integrals may be performed; and writing $M_{eik} = \delta^4(p_1+p_2-p_1'-p_2')T_{eik}$, one finds

$$\frac{\partial T_{eik}}{\partial g^2} = -i\left(\frac{2pE}{m^2}\right) \int d^2z\, e^{i\vec{q}\cdot\vec{z}} \frac{\partial}{\partial g^2}\cdot$$

$$\cdot\exp[-i \frac{g^2}{(2\pi)^2} \gamma(E)\int \frac{d^2k}{k^2+\mu^2} e^{i\vec{k}\cdot\vec{z}}],$$

or

$$T_{eik}(s,t;g^2) = \frac{is}{2m^2} \int d^2b\, e^{i\vec{q}\cdot\vec{b}}[1-e^{i\chi_o(\vec{b};s)}], \tag{9.13}$$

with the eikonal function given by

$$i\chi_o(\vec{b};s) = -\frac{ig^2}{(2\pi)^2}\gamma(s)\int\frac{d^2k}{k^2+\mu^2}e^{i\vec{k}\cdot\vec{b}} =$$

$$= -\frac{ig^2}{2\pi}\gamma(s)K_o(\mu b). \qquad (9.14)$$

The total cross section σ_T is related by the optical theorem to the CM scattering amplitude $T(s,t)$ (with normalization defined by the equations of this section) according as

$$\sigma_T = \frac{m^2}{pE} I_m T(s,t=0) \simeq \frac{4m^2}{s} I_m T(s >> m^2,\ t=0),$$

and we obtain from (9.13)

$$\sigma_T = 2\mathrm{Re}\int d^2b[1-e^{i\chi_o}]. \qquad (9.15)$$

Since $\gamma(s) \to 1$ as $s \to \infty$, χ_o and the σ_T computed from (9.15) become independent of s in the high energy limit, in apparent agreement with the observed, constant pp total cross sections. The invariant differential cross section is given by

$$\frac{d\sigma}{dt} = \frac{m^4}{16\pi p^2E^2}|T|^2 \simeq \frac{m^4}{\pi s^2}|T|^2 \qquad (9.16)$$

at high energies. Substitution of the T_{eik} of (9.13) into (9.16) produces diffraction-like forward peaks for small and moderate $-t$. There is only the barest experimental suggestion of the larger $-t$ dip structure which follows from T_{eik}; but for small $-t < 1\ (\mathrm{Gev})^2$, the agreement with observed pp scattering (considering the exchanged NVM to be the ω) is quite good.[3]

An interesting observation has been made[6] concerning the appearance of bound states in those situations where the cross-channel invariant (here, -t) is allowed to become "large," since the resulting Regge-like amplitude exhibits poles when the direct channel invariant

(s) is continued below threshold $(4m^2)$. Essentially, μ^2 is the parameter which sets the scale for the "largeness" of $-t$, and one may consider (9.13) in the kinematical region $s > 2m^2$, $4m^2 > s > -t > \mu^2$. For the possibility of binding we must, of course, consider oppositely charged fermions, and hence change g^2 to $-g^2$ in χ_o. With the standard continuation $\gamma(s) \to i\beta(s)$, $\beta(s) = (s-2m^2)[s(4m^2-s)]^{-1/2}$, and use of the asymptotic form $K_o(z) \sim -\ln Z$, it is straightforward to show that in this region

$$T_{eik} \sim -\frac{i\pi}{m^2}\left(\frac{s}{\mu^2}\right)\int_o^\infty dz \cdot z^{1-i\frac{g^2}{2\pi}\gamma(s)} J_o\left(\frac{q}{\mu}z\right),$$

or

$$\frac{2m^2}{s}T_{eik} \sim -\frac{i\pi}{\mu^2}\left(\frac{-t}{4\mu^2}\right)^{\alpha(s)-1} \cdot \frac{\Gamma\big(1-\alpha(s)\big)}{\Gamma\big(\alpha(s)\big)}, \qquad (9.17)$$

where $\alpha(s) = (g^2/4\pi)\beta(s)$. This has the Regge asymptotic form[7] for large $-t$ with trajectory function $\alpha(s)$, and displays poles whenever $\alpha(s) = n = 1, 2,\cdots$; the same ratio of Γ functions is found in the exact solution to the Coulomb scattering problem. With $s = (2m-\varepsilon)^2$, $\varepsilon << m$, this condition is equivalent to the nonrelativistic bound state (Bohr) energy levels for a pair of equal mass particles, $\varepsilon_n = (g^2/4\pi)^2 \cdot m/4n^2$. However, this procedure must physically correspond to using the Coulomb tail of a more realistic potential when solving the Schroedinger equation, and yields the correct Bohr levels because of the fortunate degeneracy of ε_n with respect to ℓ (or j). It is difficult to see how relativistic corrections to ε_n, which depend on both n and ℓ (or j) could be correctly given by this procedure.

Many phenomenological variants of (9.13) have been attempted by writing forms for $\chi(b;s)$ which yield Regge pole amplitudes, or Regge amplitudes modified by multiple scattering corrections.[8] Perhaps the most exciting set of formulae have been obtained by Cheng and Wu in a series of papers[9] concerned with summing the leading high-energy behavior of sets of Feynman graphs in a gluon field theory. Central to their analysis are the so-called tower graphs, built out of special, closed-fermion-loop contributions, which generate important s dependence in the eikonal function, in an approximation going beyond that of (9.14). It is not difficult to show that the exact T_{eik} may be written in the form (9.13) if the eikonal is given by[10]

$$\chi = \sum_{n=0}^{\infty} \chi_n(b;s) \tag{9.18}$$

where each χ_n represents the contribution of the complete, connected, cross-channel $\ell NVM \to \ell' NVM$ scattering amplitude, with $\ell + \ell' = n-2$. In terms of previous functional notation for the closed-loop-functional $L[A]$, the exact eikonal for this interaction is given by

$$i\chi(b;s) = ig^2\int f_I \Delta_c f_{II} + \bar{L}[g\int (f_I + f_{II})\Delta_c], \tag{9.19}$$

or

$$\exp \bar{L}[A] = \exp[-\frac{i}{2}\int \frac{\delta}{\delta A}\Delta_c \frac{\delta}{\delta A}]\cdot \exp L[A], \tag{9.20}$$

and all terms in the perturbation expansion of $\bar{L}$ independent of f_I or f_{II}, or containing but a single f_I and more than one f_{II}, or a single f_{II} and more than one f_I, are to be omitted. The logarithm of (9.20) represents the sum over all connected (massive) photon-photon amplitudes, including all of the latters' radiative corrections built out of the operation of $\exp[-\frac{i}{2}\int \frac{\delta}{\delta A}\Delta_c \frac{\delta}{\delta A}]$ upon $\exp L[A]$,

$i\chi =$ 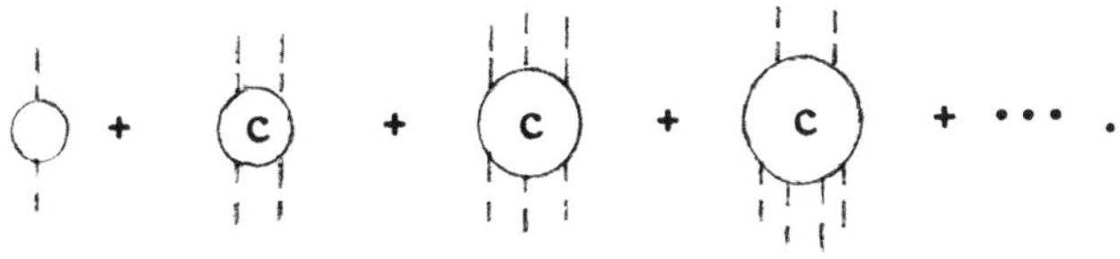 (9.21)

The subset of two-photon--two-photon amplitudes, represented by the second RHS term of (9.21), has been estimated by Cheng and Wu to produce for large impact parameter an eikonal function

$$i\chi_2(b;s) \sim -s^a \exp[-\mu_1 b], \qquad (9.22)$$

where $a \sim g^2$ and μ_1 are real constants. The eikonal of (9.22) completely dominates that of (9.14). When inserted into (9.15), with $\chi_o+\chi_2$ replacing χ_o, it produces the result

$$\sigma_T \sim (\ell n\ s)^2, \qquad (9.23)$$

saturating the Froissart bound on the possible growth of cross sections at high energy.[11] Equation (9.23) follows from (9.15) because the latter's integration cuts off at impact parameters for which $s^a \sim \exp[\mu,b]$, or $b \sim \ell n\ s$. Cheng and Wu have presented qualitative arguments which suggest that contributions from the higher connected amplitudes, defining all the remaining χ_n, do not appreciably change this σ_T (as well as their other, specific predictions). However, at present, this extremely interesting conjecture must be regarded as an open question. It is not difficult to imagine counter-examples, wherein the remaining χ_n sum together to produce s-dependence which dominates that of χ_2; one such model will be described in the next chapter.

B. Wide-Angle Approximations

One aspect of a wide angle eikonal formalism has already been mentioned, the multiplicative association of nucleon form factors with T_{eik}.[3] However, for really large momentum transfers in the limiting region $s \to \infty$, t/s finite, not all the exchanged NVM can be soft, and one must adopt a treatment in which "hard" exchanges can take place. Different authors[12] have described as many methods of performing such computations, with varying degrees of rigor; in this section, a brief, nonrigorous treatment touching on the general ideas and some moderately successful applications, will be described by yet another method designed to minimize counting arguments, while stressing the essence of all such wide-angle approximations.

The basic assumption is that scattering at large momentum transfer takes place through the action of a single hard meson exchange, which the multiple soft exchanges tend to damp away. This picture may easily be generalized to include several hard exchanges, although no estimates along these lines have as yet been performed. Single hard meson exchange, either of a neutral (for pp scattering) or appropriately charged (for np scattering) meson, represents a relativistic generalization of the Schiff formula[13] for wide-angle potential scattering, wherein a fast particle may experience many small deflections as it enters a region of nonzero potential, suffers a single large-angle scattering to essentially its final direction, and undergoes many small-angle deflections as it leaves the scattering region. The distinguishability of soft and hard exchanges, and the ensuing approximation of a general form such as (9.2) (neglecting, for simplicity, all closed-fermion loops), may be introduced in a definite, if somewhat arbitrary way, by defining

$$\tilde{\Delta}_{c,S}(k) = \frac{e^{-i\alpha k^2}}{k^2+\mu^2}, \quad \tilde{\Delta}_{c,H}(k) = \frac{1}{k^2+\mu^2}\left[1-e^{-i\alpha k^2}\right], \qquad (9.24)$$

with α a constant identified below. The purpose of this separation is to insure that only small k (more precisely, k^2) can contribute to integrals over $\tilde{\Delta}_{c,S}$, and only large k may contribute to those over $\tilde{\Delta}_{c,H}$.

With $\tilde{\Delta}_c \equiv \tilde{\Delta}_{c,S} + \tilde{\Delta}_{c,H}$, the exponential operator of (9.2) may be expanded in powers of $\tilde{\Delta}_{c,H}$,

$$M(x_1y_1,x_2y_2) \sim i^2\exp[-\frac{i}{2}\int \frac{\delta}{\delta A}\Delta_{c,S}\frac{\delta}{\delta A}]\cdot$$
$$\cdot G_I(y_1,x_1|A)G_{II}(y_2,x_2|A)\Big|_{A\to 0} +$$
$$+\left(-\frac{i}{2}\int \frac{\delta}{\delta A}\Delta_{c,H}\frac{\delta}{\delta A}\right)\cdot$$
$$\cdot i^2\exp[-\frac{i}{2}\int \frac{\delta}{\delta A}\Delta_{c,S}\frac{\delta}{\delta A}]\cdot G_I\cdot G_{II}\Big|_{A\to 0}, \tag{9.25}$$

retaining no more than a single hard exchange.

The first line of (9.25) may be evaluated in almost exactly the same manner as used in the previous section; the only differences are that $\Delta_{c,S}$ of (9.24) replaces Δ_c of (9.3), the vertex t-dependence must be included, and the replacements of (9.7) by (9.9), and $\bar{u}(p'_{1,2})\gamma_\mu u(p_{1,2})$ by $-ip^\mu_{1,2}/m$ are not to be made. This term then generates a contribution T'_{eik} given by

$$\frac{\partial}{\partial g^2}T'_{eik} = \left(\bar{u}(p'_1)\gamma_\mu u(p_1)\right)\left(\bar{u}(p'_2)\gamma_\mu u(p_2)\right)\cdot\int d^4z\ e^{iq\cdot z}\cdot$$
$$\cdot\Delta_{c,S}(z)\cdot\exp[i\chi'_o+2\phi(t)], \tag{9.26}$$

where $i\chi'_o(z;p_1,p'_1,p_2,p'_2) = i\int f^\mu_I\,\Delta_{c,S}\,f^\mu_{II}$, and

$$\phi(t) = \frac{i}{2}\frac{g^2}{(2\pi)^4}\int\frac{d^4k}{k^2+\mu^2}\,e^{-i\alpha k^2}\left(\frac{p}{k\cdot p}-\frac{p'}{k\cdot p'}\right)^2. \tag{9.27}$$

It should be noted that, in this model approximation,

the integrand of (9.27) is independent of any configuration space variable, and hence a cut off, such as that provided by α, is needed. In fact, (9.27) may be used to suggest the detailed nature of this cut off by requiring $\phi(t)$ to have the same phase for negative t as that of the finite integral containing quadratic $(k^2 \pm 2k \cdot p)$ denominators. A simple, covariant procedure which satisfies this condition consists of the two steps: (i) perform all $\int d^4k$, treating α as a real, positive number, and (ii) after all integrations have been completed, perform the continuation $\alpha \to -i\mu_c^{-2}$, where μ_c represents a real, positive cut off with dimensions of mass. With this prescription, (9.27) may easily be evaluated[4] to yield $\phi(t) = \gamma F(t)$, where

$$F(t) = t\int_{4m^2}^{\infty} \frac{dt'}{t'} \frac{1}{t'-t} \left(1- \frac{2m^2}{t'}\right)\left(1- \frac{4m^2}{t'}\right)^{-1/2} \tag{9.28}$$

and

$$\gamma = \frac{g^2}{8\pi^2}\int_0^{\infty} \frac{db}{b+\mu_c^{-2}} e^{-b\mu^2} \simeq \frac{g^2}{8\pi^2} \ln(1+\mu_c^2/\mu^2).$$

One pleasant feature of this method of approximation is that the three constants needed to specify the soft exchanges, g, μ, and μ_c, coalesce into the single constant γ, which may be treated as a parameter to be determined by experiment. For negative t, $F(t)$ is real and negative,

$$F(t) = 1- \frac{2x+1}{[x(x+1)]^{1/2}} \ln\left(\sqrt{x} + \sqrt{x+1}\right), \quad x \equiv - \frac{t}{4m^2} > 0, \tag{9.29}$$

with limiting forms $t/3m^2$ and $-\ln(-t/m^2)$ for small and large $-t$, respectively. When m denotes the nucleon mass, a very good approximation to (9.29) for all $-t > 0$ is given by $\left(|t| \text{ in } (\text{Gev})^2\right)$

$$F(t) \simeq -\ln\,(1+.4|t|) \tag{9.30}$$

Incidentally, the choice $\mu_c^2 \sim |t|$ for large $-t$ reproduces the Jackiw leading-log form of the asymptotic $\Gamma_\mu(t)$ mentioned in Chapter 8, Section C; however, we shall continue to keep μ_c a constant $\gtrsim \mu$. For positive argument, it is not difficult to show that

$$\text{Re } F(s) = F(4m^2-s), \quad s > 4m^2, \tag{9.31}$$

a relation which will subsequently be most useful.

An estimate for γ may be obtained by using the physical content of $\exp[\gamma F(t)]$ as representing the damping due to soft, virtual NVM exchange of the nucleon vertex. Hence one expects the proton's electromagnetic form factors to consist of this soft dependence multiplying an unspecified hard meson contribution $H(t)$, which may be thought of as an appropriate collection of vector mesons, represented by simple poles, and responsible for large momentum transfer processes. The crude choice $H(t) = 1+at(bm_p^2-t)^{-1}$ for the isovector form factor produces large momentum transfer damping of form

$$G(t) \sim (1-a)e^{\gamma F(t)} \sim (1-a)\left(\frac{|t|}{m^2}\right)^{-\gamma}, \tag{9.32}$$

which, with $a = 0.84$ and $\gamma = 2.4$, provides good agreement with recent SLAC data;[14] in fact, with $b \sim 2/3$, one easily reproduces the experimental curves for all $-t$. An excellent two-parameter fit at all momentum transfers is given by

$$G(t) = (1-t/m_o^2)^{-1}\, e^{\gamma F(t)},$$

with $\gamma = 1.4$ and $m_o^2 \simeq \frac{2}{3} m_p^2$; in both cases, an appropriate choice of γ leads to the experimentally observed damping, slightly in excess of that given by the empirical dipole-fit formula.

Away from the forward direction, the simplifying

operations following (9.12) may not be performed, and one is left, in (9.25), with a rather cumbersome form. However, when $-t$ is large one expects only small z to be important, and this suggests the adoption of an approximation which neglects the z dependence of $i\chi_0'$. The integrals defining χ_0' already contain the soft cut-off α, and may be evaluated in exactly the same manner as (9.28). Physically, such a dipole approximation (the neglect of z inside χ_0') decouples soft from hard effects, using the α-prescription to limit the magnitude of soft NVM exchanges, a task which should properly be accomplished by the detailed form of the hard exchanges. As a consequence of this simplifying approximation, $i\chi_0' \simeq 2\gamma[F(u)-F(s)]$, with the negative sign appearing in front of $F(s)$ because of the kinematical difference between the definitions $u = -(p_1-p_2')^2 = -(p_2-p_1')^2$ and $s = -(p_1+p_2)^2 = -(p_1'+p_2')^2$. The $\int d^4z$ of (9.26) may then be carried through, and one obtains

$$T'_{eik} \simeq g^2(\bar{u}\gamma_\mu u)_I(\bar{u}\gamma_\mu u)_{II} \frac{\exp[-|t|/\mu_c^2]}{\mu^2+|t|} \cdot$$

$$\cdot \left\{\frac{e^{2\gamma[F(t)+F(u)-F(s)]}-1}{2\gamma[F(t)+F(u)-F(s)}\right\}, \tag{9.33}$$

which, by virtue of the $\exp[-|t|/\mu_c^2]$ cut off, is negligible compared to the large $-t$ contribution arising from the second term of (9.25). In the absence of soft effects, the latter would simply correspond to the Born approximation for nucleon-nucleon scattering; with soft NVM exchange, all the above steps and approximations may be carried through and lead to the result

$$T(s,t) \simeq T_B(s,t)\ \exp[2\gamma\big(F(t)+F(u)-F(s)\big)], \tag{9.34}$$

where T_B denotes the Born approximation (suitably symmetrized for pp scattering) amplitude, with $[1-\exp(-|t|/\mu_c^2)]$ replaced by unity.

The substitution of (9.34) into (9.16) produces a $d\sigma/dt$ with several agreeable features. Because of the constraint $s+t+u = 4m^2$, and the condition (9.31), one finds the form

$$\frac{d\sigma}{dt} = \left(\frac{d\sigma}{dt}\right)_{\text{Born}} \cdot \exp[4\gamma\left(F(t)+F(4m^2-s-t)-F(4m^2-s)\right)], \quad (9.35)$$

which, with a single hard ρ_0-exchange defining $(d\sigma/dt)_B$, and the same $\gamma \simeq 2.4$ used for the nucleon electromagnetic form factor, produces a fit to recent high energy data over wide ranges of s and t which must be considered as quite good, considering the paucity of parameters and the crudeness of the calculation. For $s \to \infty$ and fixed (but large) $-t$, the RHS of (9.35) is proportional to $\exp[4\gamma F(t)] \sim \left(G(t)\right)^4$, the celebrated Chou-Yang-Wu observation.[15] For large $s,-t,-u$, these exponential factors automatically produce dependence upon the Krisch variable ut/s, and may be used to fit the large momentum transfer part of the experimental curves.[16] Finally, it may be remarked that a similar model for wide angle np scattering can serve to lift the degeneracy between soft ω and soft ρ_0 exchange, producing qualitative fits to the observed asymmetry in forward/backward np scattering at high energies.[17] It may be hoped that the qualitative success of such crude models is not completely fortuitous.

C. Inelastic (Bremsstrahlung) Models

The experimentally observed damping properties of individual scattering amplitudes may be physically understood as due to unitarity restrictions caused by the opening up of new inelastic channels. The gluon models of the previous two sections have multiple inelastic unitarity contributions of bremsstrahlung form, with massive NVMs emitted from the nucleon legs of any given hadronic process; if the NVM are identified as ρ_0 or ω, they immediately decay into the multiple pions experimentally observed. Such "fragmentation" differs conceptually from the "pionization" picture of the multiperipheral calculations, where many soft (CM) pions are emitted directly from the innards of a graph, rather than from its legs. In this section, a general (functional) formula for inclusive processes, or multiplicities, will be derived and applied in a most approximate way to bremsstrahlung emission; an application to multiperipheral processes is discussed in Chapter 10.

For definiteness, we consider inelastic pp scattering, according to the reaction $p_1+p_2 \to p_1'+p_2'+\sum k_\ell$ where $\sum k_\ell$ denotes an arbitrary number of emitted NVMs, which subsequently decay into pions. Using the notation of Chapter 3, the probability for the emission of n NVM compatible with fixed initial (p_1,p_2) and final (p_1',p_2') proton momenta is given by

$$P_n = \sum_{\gamma_n} |\langle k_1,\cdots k_n;p_1',p_2'|S|p_1,p_2\rangle|^2, \tag{9.36}$$

where $\sum_{\gamma_n}$ denotes all possible final state summations over the phase space of n NVMs; the closure statement for these particles alone would read

$$\sum_{n=0}^{\infty} \sum_{\gamma_n} |k_1,\cdots k_n\rangle\langle k_1,\cdots k_n| = 1.$$

The S-matrix of (2.25) may be rewritten in a form appropriate to the present case,

$$\frac{S}{\langle S\rangle} = :\exp[\int A_{IN} \frac{\delta}{\delta A_\mu}]:S_F\{A\}\Big|_{A\to 0}, \tag{9.37}$$

where

$$S_F\{A\} = :\exp Z_2^{-1/2}\int[\bar{\psi}_{IN}\mathcal{D} \frac{\delta}{\delta\bar{\eta}} - \frac{\delta}{\delta\eta} \bar{\overleftarrow{\mathcal{D}}}\psi_{IN}]:Z\{A,\eta,\bar{\eta}\}\Big|_{\eta\to\bar{\eta}\to 0}, \tag{9.38}$$

and $A_\mu \equiv \int \Delta_c j_\mu$. In (9.37) the $Z_3^{-1/2}$ dependence has been replaced by unity, since the self-energy structure of the NVM is not of interest here. Except for unimportant phase factors, which do not enter into P_n, (9.38) is just the S-matrix for a fermion system coupled to a fictitious, c-number, external field A_μ.

As far as the emitted NVM are concerned, one may proceed exactly as in Chapter 3, Section C to obtain an expression for P_n in the form

$$P_n = \frac{1}{n!}\left(i\int \frac{\delta}{\delta A'_\mu}\Delta_{(+)}\frac{\delta}{\delta A_\mu}\right)^n T^\dagger\{A'\}T\{A\}\Big|_{A\to A'\to 0} \tag{9.39}$$

where $T(p_1p_2p_1'p_2';A) \equiv \langle p_1',p_2'|S_F\{A\}|p_1,p_2\rangle$. In the limit $A \to 0$, $T(p_1p_2p_1'p_2';0)$ is just the elastic scattering amplitude of a pair of protons. Since P_n depends upon the incident and final proton momenta (and spins), a "differential multiplicity" for NVM production may be defined as

$$\langle\nu\rangle = \sum_{n=0}^{\infty} nP_n \Big/ \sum_{n=0}^{\infty} P_n = \frac{\partial}{\partial\lambda}\ln\left\{\exp[i\lambda\int \frac{\delta}{\delta A'}\Delta_{(+)}\frac{\delta}{\delta A}]\cdot\right.$$

$$\left.\cdot\ T^\dagger\{A'\}\cdot T\{A\}\Big|_{A=A'=0}\right\}_{\lambda=1}, \tag{9.40}$$

where $\langle\nu\rangle$ depends upon the six independent kinematical invariants formed from $p_1,p_2,p_1',p_2',\sum_{\ell=1}^{n}k_\ell$. The total NVM multiplicity may here be defined as

$$\langle n\rangle = \int(dp_1')\int(dp_2')\sum_{n=0}^{\infty}nP_n \Big/ \int(dp_1')\int(dp_2')\sum_{n=0}^{\infty}P_n, \tag{9.41}$$

where $\int(dp')$ denotes integration over final proton phase space; and very crude arguments in the context of this model may be given to obtain $\langle n \rangle \sim \ell n\ s$ as $s \to \infty$, in agreement with recent inclusive experiments.[18] Rough predictions for $\langle \nu \rangle$ have also been made,[19] but not yet tested.

Qualitatively, one can anticipate the forms which soft NVM multiplicities will take by returning to the calculation of Chapter 8, Section C, where a Poisson distribution was obtained for soft photon emission. If k_{min} is replaced by μ, and both upper cut offs ΔE and m are replaced by the maximum CM energy available $\sim \sqrt{s}$, one directly obtains

$$\langle n \rangle \sim \frac{q^2}{m^2}\ \ell n\ (s/\mu^2)$$

for small q^2/m^2; for larger momentum transfer, q^2 values of the same order as s, one would expect $\langle n \rangle \sim \ell n^2 s$, an effect recently predicted[20] in this gluon model using the technique of summing leading powers of $\ell n\ s$, or $\ell n\ q^2$. Recent experiments on deep-inelastic electron-proton scattering,[21] in which total cross sections remain large, may be understood in at least a qualitative way in the context of the same bremsstrahlung model.[20,22] However, alternate descriptions of the physics abound,[23] and detailed calculations of the predictions of all models, fitted against the rapidly expanding experimental data, will be most welcome.

Notes

1. G. Tiktopoulos and S. Treiman, Phys. Rev. D2, 805 (1970).

2. A qualitative description of why and when eikonalization of simple exchanges is valid, in terms of the possible routings of large momenta through different propagators in various Feynman graphs, has been given by E. Eichten and R. Jackiw, Phys. Rev. D4, 439 (1971).

3. Y-P. Yao, Phys. Rev. D2, 1342 (1970).

4. H. M. Fried and T. K. Gaisser, Phys. Rev. 179, 1491 (1969).

5. T. K. Gaisser, Phys. Rev. D2, 1337 (1970).

6. M. Lévy and J. Sucher, Phys. Rev. D2, 1716 (1970); see also E. Brezin, C. Itzykson, and J. Zinn-Justin, Phys. Rev. D1, 2349 (1970).

7. A recent summary of the Regge formalism has been given by V. D. Barger and D. B. Cline, *Phenomenological Theories of High Energy Scattering*, W. A. Benjamin, Inc., New York (1969).

8. For example, S. Frautschi and B. Margolis, Nuovo Cimento 57A, 427 (1968); R. C. Arnold, Phys. Rev. 153, 1523 (1967).

9. A recent paper containing sets of predictions is that of H. Cheng and T. T. Wu, Phys. Rev. Letters 24, 1456 (1970).

10. H. M. Fried, Phys. Rev. D3, 2010 (1971). In the absence of closed-fermion loops, the techniques of this reference can provide an alternate derivation of (9.13). The passage from (9.5) to (9.6) is permissible in the $t/s \to 0$ limit, but is incomplete in the wide-angle situation; however, the conclusion leading to (9.34) remains valid.

11. M. Froissart, Phys. Rev. 123, 1053 (1961).

12. R. Blankenbecler and R. L. Sugar, Phys. Rev. D2, 3024 (1970); H. D. I. Abarbanel and C. Itzykson, Phys. Rev. Letters 23, 53 (1969); J. Cardy, Nucl. Phys. B17, 493 (1970); and Refs. 4 and 5 of this chapter.

13. L. I. Schiff, Phys. Rev. 103, 443 (1956).

14. D. H. Coward, *et al.*, Phys. Rev. Letters 20, 292 (1968).

15. T. T. Wu and C. N. Yang, Phys. Rev. 137, B708 (1965); and T. T. Chou and C. N. Yang, Phys. Rev. Letters 20, 1213 (1968).

16. C. W. Akerlof, *et al.*, Phys. Rev. 159, 1138 (1967).

17. H. M. Fried and K. Raman, Phys. Rev. D3, 269 (1971).

18. L. W. Jones, *et al.*, Phys. Rev. Letters 25, 1679 (1970).

19. H. M. Fried and T. K. Gaisser, Phys. Rev. D4, 3330 (1971).

20. P. M. Fishbane and J. D. Sullivan, Physics Letters 37B, 68 (1971).

21. R. Taylor, Proceedings of the International Symposium on Electrons and Photons at High Energies, Liverpool, England, 1969, edited by W. D. Braben (Daresbury Nuclear Physics Laboratory, 1970).

22. H. M. Fried and H. Moreno, Phys. Rev. Letters 25, 625 (1970).

23. For example, S. Drell and T-M. Yan, Phys. Rev. Letters 24, 181 (1970); J. D. Bjorken, Phys. Rev. 179, 1547 (1969); G. B. West, Phys. Rev. Letters 24, 1206 (1970).

Chapter 10

SPECULATIONS AT HIGH ENERGY

This chapter deals with the estimation of an amplitude's high energy behavior by the technique of summing its leading s-dependence in every order of perturbation theory. Such a mathematically ill-defined procedure carries with it the tacit assumption that the sum of every order's next-to-leading behavior is of less importance, as well as the more obvious expectation that the sum of the leading s-dependence does correspond to the true high-energy limit. The modified A^3 theory treated here provides a good example of the use of functional methods in identifying the complete eikonal of the problem, and when combined with a method of estimating the leading $\ell n\ s$ terms of contributing graphs, reproduces those results of existing calculations based upon the s-channel iteration of tower graphs. The final section deals with a sequence of speculations which try to estimate and sum the leading log behavior of every graph contributing to the eikonal; what emerges is the possibility of multiple cancellations destroying the basis of (9.23), which may be indicative of the true high energy limits in field theory. Although the correct answer is unknown, these methods at least provide a formalism within which various approximations may be developed and explored.

A. Multiperipheral Field Theory

Although the interaction $L' = -gA^3$ provides the simplest, nontrivial structure, it has several disadvantages among which is the inability of scalar mesons to eikonalize properly. We shall begin this section by writing the complete generating functional for a more complicated interaction,

$$L' = -ig\bar{\psi}\Big(\sum_{\mu} \gamma_{\mu}\cdot w_{\mu}\Big)\psi - \frac{\lambda}{2}\pi \sum_{\mu} w_{\mu}^{2}, \tag{10.1}$$

containing a nucleon field ψ, a NVM field w_{μ}, and a scalar "pion" field π; and then discarding, in the functional expression for nucleon-nucleon scattering, those subsets of graphs which contain ψ, w_μ, and π self-interactions, until all that remains are graphs containing multiple NVM exchange between nucleons, with scalar pion exchange taking place between the virtual NVM. In analogy with previous forms, the NVM interaction may be expected to eikonalize, carrying with it the composite substructures here defined by multiple pion exchange between different w_μ. This corresponds to the possible graphs of A^3 theory, as well as to the closed loop contributions of massive-photon QED.

The generating functional for the interaction (10.1) is given by

$$Z\{j,k_\mu,\eta,\bar{\eta}\} = <\left(\exp i\int [j\pi+k\cdot w+\bar{\eta}\psi+\bar{\psi}\eta]\right)_+>,$$

where $j,k_\mu,\eta,\bar{\eta}$ denote c-number sources for the fields $\pi,w_\mu,\bar{\psi},\psi$, respectively. Following the techniques of Chapter 3, the formal solution for Z is

$$<S>Z = \exp[-i\int \frac{\delta}{\delta\eta}\left(g\gamma\cdot \frac{\delta}{\delta k}\right)\frac{\delta}{\delta\bar{\eta}} - \frac{i}{2}\int \frac{\delta}{\delta k}\cdot\left(-\frac{\lambda}{i}\frac{\delta}{\delta j}\right)\frac{\delta}{\delta k}]\cdot$$

$$\cdot\exp[i\int\bar{\eta}\, S_c\eta + \frac{i}{2}\int k\cdot\Delta_c k + \frac{i}{2}\int jD_c j], \qquad (10.2)$$

with propagators $\delta_{\mu\nu}\Delta_c$ for the NVM field w, and D_c for the pion (of small, but nonzero mass μ) field π. Using the very convenient combinatorics of Chapter 3, Section D, (10.2) may be rewritten as

$$<S>Z = \exp[i\int\bar{\eta}\left(S_c[1+g\frac{\delta}{\delta k}\cdot\gamma S_c]^{-1}\right)\eta+\mathrm{Tr}\ln\left(1+g\gamma\cdot\frac{\delta}{\delta k}S_c\right)]\cdot$$

$$\cdot\exp[\frac{i}{2}\int k\cdot\left(\Delta_c[1+\frac{\lambda}{i}\frac{\delta}{\delta j}\Delta_c]^{-1}\right)k-\frac{1}{2}\mathrm{Tr}\ln\left(1+\frac{\lambda}{i}\frac{\delta}{\delta j}\Delta_c\right)]\cdot$$

$$\cdot\exp[\frac{i}{2}\int jD_c j], \qquad (10.3)$$

while the neglect of all closed nucleon and closed NVM loops yields the simpler quantity

$$Z \simeq \exp[i\int \bar{\eta}\left(S_c[1+g\gamma\cdot \frac{\delta}{\delta k}\, S_c]^{-1}\right)\eta]\cdot$$

$$\cdot\exp[\frac{i}{2}\int k\cdot\Delta_c[1+\frac{\lambda}{i}\frac{\delta}{\delta j}\,\Delta_c]^{-1}k]\cdot$$

$$\cdot\exp[\frac{i}{2}\int jD_c j]. \qquad (10.4)$$

The configuration space nucleon-nucleon scattering amplitude may be obtained, as in (9.2), by the appropriate functional differentiation of (10.4),

$$M(x_1y_1,x_2y_2) = i^2 G_I\left(y_1x_1\middle|\frac{g}{i}\frac{\delta}{\delta k}\right)G_{II}\left(y_2x_2\middle|\frac{g}{i}\frac{\delta}{\delta k}\right)\cdot$$

$$\cdot\exp[\frac{i}{2}\int k\cdot\Delta_c[1+\frac{\lambda}{i}\frac{\delta}{\delta j}\,\Delta_c]^{-1}k]\cdot$$

$$\cdot\exp[\frac{i}{2}\int jD_c j]\Bigg|_{\substack{k_\mu\to 0,\\ j\to 0,}} \qquad (10.5)$$

where the nucleons are again considered distinguishable, to avoid symmetrization. A slightly more convenient form of (10.5) is

$$M = i^2 \exp[-\frac{i}{2}\int \frac{\delta}{\delta k}\cdot\Delta_c[1+\frac{\lambda}{i}\frac{\delta}{\delta j}\,\Delta_c]^{-1}\frac{\delta}{\delta k}]\cdot$$

$$\cdot G_I(y_1x_1|gk)G_{II}(y_2x_2|gk)\cdot\exp[\frac{i}{2}\int jD_c j]\Bigg|_{k_\mu\to 0,\, j\to 0,} \qquad (10.6)$$

and we next omit all graphs containing W_μ lines emitted and absorbed by the same nucleon, retaining only the exchange of virtual NVMs between different fermions,

$$M \to i^2\exp[-i\int \frac{\delta}{\delta k_1}\cdot\Delta_c[1+\frac{\lambda}{i}\frac{\delta}{\delta j}\,\Delta_c]^{-1}\frac{\delta}{\delta k_2}]\cdot G_i(y_1x_1|gk_1)\cdot$$

$$\cdot G_{II}(y_2x_2|gk_2)\cdot \exp[\tfrac{i}{2}\int jD_cj]\Big|_{k^\mu_{1,2}\to 0,\ j\to 0}. \quad (10.7)$$

As far as the NVM $k^\mu_{1,2}$ source dependence is concerned, (10.7) is completely analogous to (9.2), with

$$\Delta_c[1+\frac{\lambda}{i}\frac{\delta}{\delta j}\Delta_c]^{-1} \quad \text{and} \quad \exp[\tfrac{i}{2}\int jD_cj]$$

replacing Δ_c and $L[A]$, respectively. An eikonal calculation may now be initiated, in the $s\to\infty$, $t/s\to 0$ limit, by extracting the soft W-meson dependence of each $G[k]$; in place of (9.6) one has

$$\frac{\partial M}{\partial g^2} = -\left(\frac{p_1\cdot p_2}{m^2}\right)(2\pi)^{-4}\int d^4z_1\ e^{iz_1\cdot(p_1-p_1')}\cdot$$

$$\cdot\int d^4z_2\ e^{iz_2\cdot(p_2-p_2')}\bar{\Delta}_c\left(z_1,z_2|\frac{\lambda}{i}\frac{\delta}{\delta j}\right)\cdot$$

$$\cdot\exp[ig^2\int d^4u\int d^4v\ \sum_\mu F^\mu_I(u)\bar{\Delta}_c\left(u,v\middle|\frac{\lambda}{i}\frac{\delta}{\delta j}\right)F^\mu_{II}(v)]\cdot$$

$$\cdot\exp[\tfrac{i}{2}\int jD_cj]\Big|_{j\to 0}, \quad (10.8)$$

where

$$\bar{\Delta}_c(x,y|\lambda\pi) = \langle x|\Delta_c[1+\lambda\pi\Delta_c]^{-1}|y\rangle,$$

$$gF^\mu_{I,II} = f^\mu_{I,II},$$

with the $f^\mu_{I,II}$ given by (9.7) and (9.9). This expression may be integrated,

$$M[g^2;\pi] = \left(\frac{is}{2m^2}\right)(2\pi)^{-4}\int d^4z_1\, e^{iz_1\cdot q_1}\int d^4z_2\, e^{iz_2\cdot q_2}\cdot$$

$$\cdot\exp\left[-\frac{i}{2}\int\frac{\delta}{\delta\pi}D_c\frac{\delta}{\delta\pi}\right]\cdot\bar{\Delta}_c(z_1,z_2|\lambda\pi)\cdot$$

$$\cdot\left[\int F_I\cdot\bar{\Delta}_c[\lambda\pi]F_{II}\right]^{-1}\left\{\exp\left[ig^2\int F_I\cdot\bar{\Delta}_c[\lambda\pi]F_{II}\right]-1\right\},$$

$$(10.9)$$

where $\pi \equiv \int D_c j$, and a factor $\exp[\frac{i}{2}\int jD_c j]$ has been omitted from the RHS of (10.9), since it contributes nothing to the elastic scattering amplitude, nor to subsequent inelastic cross sections; as before, $q_{1,2} = p_{1,2}-p'_{1,2}$, and $M(g^2) = M[g^2;\pi]\big|_{\pi\to 0}$.

To obtain from (10.9) the standard eikonal form, one notes that

$$\int F_I\cdot\bar{\Delta}_c[\pi]\cdot F_{II} = (p_1\cdot p_2)\int_{-\infty}^{+\infty}da\int_{-\infty}^{+\infty}db\ \bar{\Delta}_c(z_1-ap_1,z_2-bp_2|\pi).$$

As long as $\pi \neq 0$, this depends upon both variables separately; but because of the parametric $\underline{a,b}$ integration, the double Fourier representation has the form

$$(2\pi)^2(p_1\cdot p_2)\int d^4K\int d^4K'\ e^{iK\cdot z_1+iK'\cdot z_2}\cdot$$

$$\cdot\delta(K\cdot p_1)\delta(K'\cdot p_2)\tilde{\bar{\Delta}}_c(K,K'|\pi). \qquad (10.10)$$

In the CM, with $\vec{p}_1 = -\vec{p}_2$ again chosen along the z axis, one may write

$$k\cdot z = \vec{k}\cdot\vec{z}_T+k_3z_3-k_oz_o = \vec{k}\cdot\vec{z}+\frac{1}{2}k^{(+)}z^{(-)}+\frac{1}{2}k^{(-)}z^{(+)},$$

with $a^{(\pm)} = a_3 \pm a_o$, and it is then clear that (10.10) is independent of $z_1^{(+)}$ and $z_2^{(-)}$. All dependence on this latter pair of variables lies in the $\bar{\Delta}_c(z_1,z_2|\pi)$ term of (10.9); and hence the assumption that $q_1^{(-)} = q_2^{(+)} = 0$ in the $s \to \infty$ limit permits integration over the $z_1^{(+)}$, $z_2^{(-)}$ variables, yielding $\left(\int d^4z = \frac{1}{2}\int d^2z \int dz^{(+)} \int dz^{(-)}\right)$

$$\frac{1}{2}\int dz_1^{(+)}\int dz_2^{(-)}\bar{\Delta}_c(z_1,z_2|\pi) = -\int F_I\cdot\bar{\Delta}_c[\pi]\cdot F_{II},$$

and thereby cancelling the denominator factor of (10.9). Thus one finds the eikonal form

$$M[g^2;\pi] = \left(\frac{is}{2m^2}\right)(2\pi)^{-4}\int d^2z_1 e^{i\vec{q}_1\cdot\vec{z}_1}\int d^2z_2 e^{i\vec{q}_2\cdot\vec{z}_2}\cdot$$

$$\cdot\frac{1}{2}\int dz_2^{(+)} e^{\frac{i}{2}q_2^{(-)}z_2^{(+)}}\cdot\int dz_1^{(-)} e^{\frac{i}{2}q_1^{(+)}z_1^{(-)}}\cdot$$

$$\cdot\left\{1-\exp\left[-\frac{i}{2}\int\frac{\delta}{\delta\pi}D_c\frac{\delta}{\delta\pi}\right]\cdot\exp\left[ig^2\int F_I\cdot\bar{\Delta}_c[\pi]F_{II}\right]\right\},$$

(10.11)

with $M_{eik} = M[g^2;\pi]\Big|_{\pi\to 0}$.

That portion of the total cross section due to pion emission may be obtained, at least formally, by applying (9.39) to $M[g^2;\pi]$, with the latter replacing $T[A]$. For the present elastic scattering discussion, the c-number source π is to vanish after all functional differentiations of (10.11) have been performed. In this limit it is not difficult to show that the integrand of (10.11) depends upon the transverse combination $\vec{z}_1-\vec{z}_2$ only; hence the $z_2^{(+)}$, $z_1^{(-)}$, and $\vec{z}_1+\vec{z}_2$ integrals may be performed, yielding 4-momentum conservation,

$$\delta(\vec{q}_1+\vec{q}_2)\cdot 2\delta(q_2^{(-)})\delta(q_1^{(+)})$$

$$= \delta(\vec{q}_1+\vec{q}_2)\cdot 2\delta(q_1^{(-)}+q_2^{(-)})\cdot\delta(q_1^{(+)}+q_2^{(+)}) = \delta^4(p_1+p_2-p_1'-p_2'),$$

under the previous assumption that $q_1^{(-)} = q_2^{(+)} = 0$. With $M_{eik} = \delta^4(q_1+q_2)T_{eik}$, one obtains the standard eikonal form (9.13), with the latter's $i\chi_o$ replaced by

$$i\chi = \Big(\exp[-\frac{i}{2}\int \frac{\delta}{\delta\pi} D_c \frac{\delta}{\delta\pi}]\cdot$$

$$\cdot\exp[ig^2\int F_I\cdot\bar{\Delta}_c[\pi]F_{II}]\Big)_{\substack{\pi\to 0,\\ \text{connected}}} -1, \tag{10.12}$$

or, equivalently,

$$e^{i\chi} = \exp[-\frac{i}{2}\int \frac{\delta}{\delta\pi} D_c \frac{\delta}{\delta\pi}]\cdot\exp[ig^2\int F_I\cdot\bar{\Delta}_c[\pi]F_{II}]\Big|_{\pi\to 0}. \tag{10.13}$$

The expansion of (10.12) in powers of g^2 generates an infinite sum of contributions to the eikonal, consisting of all the connected, $nW \to nW$ t-channel amplitudes; that is, $i\chi = \sum_{n=1}^{\infty} i\chi_n$, with

$$i\chi_n = \frac{(ig^2)^n}{n!}\int F_I^{\mu_1}(u_1)\cdots F_I^{\mu_n}(u_n)M_{n,n}^{(c)}(u_1,\cdots u_n;v_1,\cdots v_n)\cdot$$

$$\cdot F_{II}^{\mu_1}(v_1)\cdots F_{II}^{\mu_n}(v_n), \tag{10.14}$$

and

$$M^{(c)}_{n,n} = \Big(\exp[-\frac{i}{2}\int \frac{\delta}{\delta\pi} D_c \frac{\delta}{\delta\pi}] \cdot$$

$$\cdot \bar{\Delta}_c(u_1,v_1|\pi)\cdots\bar{\Delta}_c(u_n,v_n|\pi)\Big|_{\pi\to 0}\Big)_{\text{conn.}} \qquad (10.15)$$

When all self-linkages of any $\bar{\Delta}_c(u,v|\pi)$, in (10.15), are omitted, and only linkages of one $\bar{\Delta}_c$ with another are retained, (10.14) provides an eikonal structure analogous to that of massive photon QED and the simpler A^3 theory. The total cross section for n-n scattering is given by (9.15), with (10.12) replacing $i\chi_o$, and is built out of the production amplitudes for the emission of W_μ and π mesons. In particular, those processes in which only pions are emitted here take on the multiperipheral form: pions emitted in sequence from an internal, t-channel (NVM) line as $s \to \infty$. They are represented in this model by every term of χ except χ_1, with the latter (when its self-energy pion structure is dropped) just reproducing the NVM χ_o of (9.14).

B. Summing Leading Logs

In the context of this example, a technique is now outlined for identifying the leading $\ell n\ s$ dependence of every Feynman graph contributing to χ. It is simplest not to expand in terms of connected graphs, but to use (10.13) directly,

$$e^{i\chi}-1 = \sum_{n=1}^{\infty} \frac{(ig^2)^n}{n!} \exp[-\frac{i}{2}\int \frac{\delta}{\delta\pi} D_c \frac{\delta}{\delta\pi}] \cdot$$

$$\cdot \int F_I^{\mu_1}(u_1)\cdots F_I^{\mu_n}(u_n)\bar{\Delta}_c(u_1,v_1|\pi)\cdots\bar{\Delta}_c(u_n,v_n|\pi)\cdot$$

$$\cdot F_{II}^{\mu_1}(v_1) F_{II}^{\mu_n}(v_n)\Big|_{\pi\to 0}, \qquad (10.16)$$

or, upon inserting (9.9),

$$\sum_{n=1}^{\infty} \frac{[ig^2(p_1 \cdot p_2)]^n}{n!} \int_{-\infty}^{+\infty} da_1 \cdots da_n \int_{-\infty}^{+\infty} db_1 \cdots db_n \cdot$$

$$\cdot \exp[-\frac{i}{2} \int \frac{\delta}{\delta\pi} D_c \frac{\delta}{\delta\pi}] \cdot$$

$$\cdot \bar{\Delta}_c(z_1 - a_1p_1, z_2 - b_1p_2) \cdots \bar{\Delta}_c(z_1 - a_np_1, z_2 - b_np_2) \Big|_{\pi \to 0} \cdot \tag{10.17}$$

As in Chapter 8, Section A, each $\bar{\Delta}_c(u,v|\pi)$ has the exact representation

$$\bar{\Delta}_c(u,v|\pi) = i(2\pi)^{-4} \int d^4q \; e^{iq\cdot(u-v)} \int_0^{\infty} d\xi \; e^{-i\xi(m^2+q^2)} \cdot$$

$$\cdot G(\xi;q;v), \tag{10.18}$$

where G satisfies the integral equation

$$G(\xi;q;v) = 1 - i\lambda \int_0^{\xi} d\xi' \int d^4k \tilde{\pi}(-k) e^{-ik\cdot v - i\xi'(k^2+2k\cdot q)} \cdot$$

$$\cdot G(\xi';q+k;v). \tag{10.19}$$

Thus, with $\int d^4q = \int d^2q \cdot \frac{1}{2} \int dq^{(+)} \int dq^{(-)}$ and the same CM coordinate system, a typical pair of parametric integrals yields

$$\iint_{-\infty}^{+\infty} da\,db\,\bar{\Delta}_c(z_1 - ap_1, z_2 - bp_2|\pi)$$

$$= \frac{i}{2E} (2\pi)^{-3} \int d^2q e^{i\vec{q}\cdot\vec{z}} \int dq^{(+)} e^{\frac{i}{2} q^{(+)} z^{(-)}} \int db e^{-ibEq^{(+)}} \cdot$$

$$\cdot\int_0^\infty d\xi e^{-i\xi(m^2+\vec{q}^2)}G(\xi;q;z_2-bp_2)\Big|_{q^{(-)}=0}, \qquad (10.20)$$

where again $z = z_1-z_2$. Translational invariance produces terms independent of the $z^{(-)}$ and z_2 variables appearing in (10.20), an effect which becomes apparent when different factors of the form (10.20) are combined to build the entire amplitude; and hence we shall simply omit this $z^{(-)}$ and z_2 dependence.

The *mth* iterate of (10.19) is

$$G^{(m)}(\xi;q;-bp_2)$$

$$= (-i\lambda)^m\cdot\int_0^\xi d\xi_1\int_0^{\xi_1}d\xi_2\cdots\int_0^{\xi_{m-1}}d\xi_m\int d^4k_1\tilde{\pi}(-k_1)\cdots\int d^4k_m\tilde{\pi}(-k_m)\cdot$$

$$\cdot\exp[ibp_2\cdot\sum_{\ell=1}^{m}k_\ell]\cdot\exp\Big[-i\xi_1(k_1^2+2q\cdot k_1)-i\xi_2(k_2^2+2k_2\cdot[q+k_1])+$$

$$+\cdots-i\xi_m(k_m^2+2k_m\cdot[q+\sum_{\ell=1}^{m-1}k_\ell])\Big], \qquad (10.21)$$

and the integrals

$$\int_{-\infty}^{+\infty}dq^{(+)}\int_{-\infty}^{+\infty}db\, e^{-ibEq^{(+)}}$$

of (10.20) over $G^{(m)}$ produce the factor

$$\frac{2\pi}{E}\int dq^{(+)}\delta\Big(q^{(+)}+\sum_{\ell=1}^{m}k_\ell^{(+)}\Big).$$

In this way, the exponential terms of (10.21) containing $q^{(+)}$ may be rewritten in terms of the $\sum_\ell k_\ell^{(+)}$. Since $q_{(-)} = 0$ already (by the $\int da$ integration), and $k^2+\mu^2 = \vec{k}_T^2+\mu^2+k^{(+)}k^{(-)}$, we have

$$\int dq^{(+)}\int db e^{-ibEq^{(+)}} G^{(m)}$$

$$= (-i\lambda)^m\left(\frac{2\pi}{E}\right)\int_o^\xi d\xi_1\cdots\int_o^{\xi_{m-1}} d\xi_m\cdot\int d^2k_1\cdots\int d^2k_m\cdot$$

$$\cdot\exp\left(-i\xi_1[\vec{k}_1^2+2\vec{q}\cdot\vec{k}_1]\right)\cdots\exp\left(-i\xi_m[\vec{k}_m^2+2\vec{k}_m\cdot(\vec{q}+\sum_{\ell=1}^{m-1}\vec{k}_\ell)]\right)\cdot$$

$$\cdot\frac{1}{2}\int dk_1^{(+)}\cdots\frac{1}{2}\int dk_m^{(+)}\cdot\int dk_1^{(-)}\cdots\int dk_m^{(-)}\cdot\tilde{\pi}(-k_1)\cdots\tilde{\pi}(-k_m)\cdot$$

$$\cdot\exp\left(-i\xi_1[k_1^{(+)}k_1^{(-)}-k_1^{(-)}\sum_{\ell=1}^m k_\ell^{(+)}]-i\xi_2[k_2^{(+)}k_2^{(-)}+k_2^{(+)}k_1^{(-)}+\right.$$

$$-k_2^{(-)}\sum_{\ell=2}^m k_\ell^{(+)}]\Big)\cdots\exp\left(-i\xi_m[k_m^{(+)}k_m^{(-)}+k_m^{(+)}\sum_{\ell=1}^{m-1}k_\ell^{(-)}+\right.$$

$$\left.-k_m^{(-)}k_m^{(+)}]\right). \tag{10.22}$$

There is now a lovely cancellation within the last group of factors of (10.22): every term of form $\exp[-i\xi_\ell k_\ell^{(+)}k_\ell^{(-)}]$ is removed, leaving only differences of the ξ_i, in the form

$$\exp[i\sum_{i<j=1}^m(\xi_i-\xi_j)k_i^{(-)}k_j^{(+)}].$$

Because the ξ_i are ordered, each $\xi_{ij}\equiv\xi_i-\xi_j$ is a positive number; and there then follows a compact way of determining the leading $\ln s$ dependence of every graph built out of these terms.

The simplest situation is that of two NVM W-lines (labeled $q_{1,2}$) linked together by m pion lines. This is represented by the functional operations joining $G^{(m)}(\xi;q_1;-b_1p_2)$ with $G^{(m)}(\eta;q_2;-b_2p_2)$; and after performing the $\int da_1 \int da_2 \cdot \int db_1 \int db_2$ as above, one obtains the quantity

$$\frac{1}{m!}\left(-i\int \frac{\delta}{\delta\pi} D_c \frac{\delta}{\delta\pi'}\right)^m \cdot (-i\lambda)^{2m}\int d^4k_1 \tilde{\pi}(-k_1)\cdots\int d^4k_m \tilde{\pi}(-k_m)\cdot$$

$$\cdot\int d^4K_1 \tilde{\pi}'(-K_1)\cdots\int d^4K_m \tilde{\pi}'(-K_m)\cdot\int_0^{\xi} d\xi_1\cdots\int_0^{\xi_{m-1}} d\xi_m\cdot$$

$$\cdot\int_0^{\eta} d\eta_1\cdots\int_0^{\eta_{m-1}} d\eta_m\cdots \exp\left(i\sum_{i<j=1}^{m}[\xi_{ij}k_i^{(-)}k_j^{(+)}+\eta_{ij}K_i^{(-)}K_j^{(+)}]\right), \tag{10.23}$$

where all dependence upon the transverse $\vec{k}_\ell$, $\vec{K}_\ell$, $\vec{q}_{1,2}$ variables has been suppressed. The functional operations of (10.23)--which avoid self-linkages--produce $m!$ topologically distinct graphs, each of which is repeated $m!$ times, corresponding to the different ways of labelling the pion momenta k_ℓ. Thus the $1/m!$ of (10.23) is removed, and one finds just the sum of $m!$ distinct Feynman graphs. For clarity, we adopt the "natural" labeling, in which k_ℓ and ξ_ℓ have the same subscript. The ladder graphs are then obtained, for any m, by the associations $k_\ell \leftrightarrow -K_\ell$, $1 \le \ell \le m$, together with the appropriate propagators arising from the functional differentiation,

$$\int d^4k_1[k_1^2+\mu^2-i\varepsilon]^{-1}\cdots\int d^4k_m[k_m^2+\mu^2-i\varepsilon]^{-1}.$$

Calling $\omega_\ell^2 = \vec{k}_\ell^2+\mu^2$, $\vec{k}^2 = k_x^2+k_y^2$, the integrals generating the $\ell n\ s$ dependence of the ladders are

$$I_m = \int_{-\infty}^{+\infty} \frac{dk_1^{(+)}}{k_1^{(+)}} \cdots \int_{-\infty}^{+\infty} \frac{dk_m^{(+)}}{k_m^{(+)}} \cdot \int_{-\infty}^{+\infty} dk_1^{(-)} [k_1^{(-)} + \frac{\omega_1^2}{k_1^{(+)}} - i\varepsilon(k_1^{(+)})]^{-1} \cdots$$

$$\cdot \int_{-\infty}^{+\infty} dk_m^{(-)} [k_m^{(-)} + \frac{\omega_m^2}{k_m^{(+)}} - i\varepsilon(k_m^{(+)})]^{-1} \cdot$$

$$\cdot \exp[i \sum_{i<j}^{m} (\xi_{ij}+\eta_{ij}) k_i^{(-)} k_j^{(+)}], \tag{10.24}$$

while all other (crossed) graphs are given by similar terms with permuted $k_i^{(-)}$, $k_j^{(+)}$ coordinates. It will be most useful to consider four simple graphs,

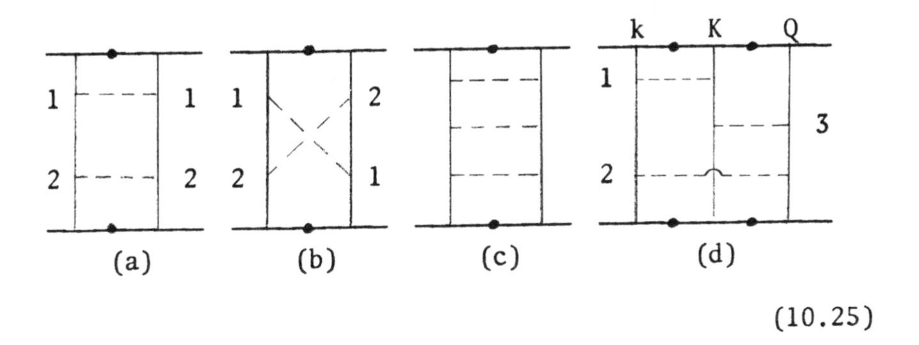

(10.25)

in brief but sufficient detail to illustrate the general pattern. The dots on the internal nucleon lines are a reminder that the eikonal limits have already been assumed; a more appropriate way of drawing these eikonal graphs would be to collapse every nucleon line so dotted into a point, as discussed in the next section.

For case (a), the integrals of form (10.24) are

$$I_2 = \int_{-\infty}^{+\infty} \frac{dk_1^{(+)}}{k_1^{(+)}} \int_{-\infty}^{+\infty} \frac{dk_2^{(+)}}{k_2^{(+)}} \int_{-\infty}^{+\infty} dk_1^{(-)} [k_1^{(-)} + \frac{\omega_1^2}{k_1^{(+)}} - i\varepsilon(k_1^{(+)})]^{-1} \cdot$$

$$\cdot \int_{-\infty}^{+\infty} dk_2^{(-)} [k_2^{(-)} + \frac{\omega_2^2}{k_2^{(+)}} - i\varepsilon(k_2^{(+)})]^{-1} \cdot$$

$$\cdot \exp[i(\xi_{12}+\eta_{12})k_1^{(-)}k_2^{(+)}, \tag{10.26}$$

or

$$I_2 = (2\pi i)^2 \int_0^{\infty} \frac{dk_1^{(+)}}{k_1^{(+)}} \int_0^{\infty} \frac{dk_2^{(+)}}{k_2^{(+)}} \exp[-i(\xi_{12}+\eta_{12})\omega_1^2 k_2^{(+)}/k_1^{(+)}. \tag{10.27}$$

The upper and lower limits of the $k_{1,2}^{(+)}$ integrals should really be replaced by $2\sqrt{s}$ and $\omega_{1,2}^2/2\sqrt{s}$, respectively, forms obtained by assuming that each $|k_3|_{max} \sim \sqrt{s}$ independently of the other. In this calculation it will also be assumed that the transverse $\vec{k}_\ell^2$ are never large, $\vec{k}_\ell^2 \lesssim \mu^2 << s$, and each ω_ℓ^2 simply replaced by μ^2. Thus, (10.27) contributes

$$I_2 = (2\pi i)^2 \int_{\frac{\mu^2}{4s}}^{1} \frac{dk_1}{k_1} \int_{\frac{\mu^2}{4s}}^{1} \frac{dk_2}{k_2} \exp[-i(\xi_{12}+\eta_{12})\mu^2 \cdot k_2/k_1], \tag{10.28}$$

where a factor $2\sqrt{s}$ has been scaled out of each $k_{1,2}^{(+)} \equiv 2\sqrt{s}\cdot k_{1,2}$. As $s \to \infty$, the leading s-dependence of (10.28) arises from that portion of the integrals which would be logarithmically divergent, were the lower limits allowed to vanish; that is, from small $k_{1,2}$. However, the phase factor $\exp[-i(\xi_{12}+\eta_{12})k_2/k_1]$ will, by rapid oscillation, remove any contribution for $k_1 \to 0$, k_2 fixed, an effect which occurs because the argument of the exponent is always of the same sign, $\xi_{12} > 0$, $\eta_{12} > 0$. More precisely, all ξ_i, η_j variables are subject to the continuation $\xi_i \to -i\tau_i$, $\eta_j \to -i\tau_j'$ performed during the final ξ_i, η_j integrations, so that an infinitely rapid phase oscillation in (10.28) does actually correspond to a zero. The only relevant part of (10.28) is that of the restricted range $k_2 < k_1$, in which case the exponential factor may be neglected,

$$I_2 \simeq (2\pi i)^2 \int_{\frac{\mu^2}{4s}}^{1} \frac{dk_1}{k_1} \int_{\frac{\mu^2}{4s}}^{k_1} \frac{dk_2}{k_2} = (2\pi i)^2 \cdot \frac{1}{2!} \left\{ \ln\left(\frac{4s}{\mu^2}\right)\right\}^2 .$$

Frequently, such forms are written as nested integrals over the "rapidity" variables $y_\ell \simeq \ln(k_\ell^{(+)}/\mu)$.

In contrast, the crossed graph (b) corresponds to the associations $k_1 \leftrightarrow -K_2$, $k_2 \leftrightarrow -K_1$, replacing the exponential factor of (10.26) by

$$\exp[i\xi_{12}k_1^{(-)}k_2^{(+)} + i\eta_{12}k_2^{(-)}k_1^{(+)}].$$

Integration over $k_{1,2}^{(-)}$ then produces a term proportional to

$$\int_o^\infty \frac{dk_1^{(+)}}{k_1^{(+)}} \int_o^\infty \frac{dk_2^{(+)}}{k_2^{(+)}} \exp\left[-i\left\{\xi_{12}\omega_1^2 \frac{k_2^{(+)}}{k_1^{(+)}} + \eta_{12}\omega_2^2 \frac{k_1^{(+)}}{k_2^{(+)}}\right\}\right], \tag{10.29}$$

which should be compared with (10.27). Again, the phase always has the same sign, but there is no log divergence at either the upper or lower limit of one of the integrals of (10.29); inserting limits, one sees that there is here but a single $\ln s$ factor, in contrast to the $(\ln s)^2$ of the two-rung ladder graph.

Pictorially, one can determine the leading log dependence of any graph by inspecting these phases, which depend upon the position of each $k_\ell^{(+)}$ on both vertical NVM lines; the maximum $\ln s$ dependence demands that neighboring rapidities be ordered. Thus (a) of (10.25) can satisfy $1 > 2$ on each vertical line, but these ordering restrictions clash for (b), which is then lower by one factor of $\ln s$. For example, the leading log dependence of the graphs

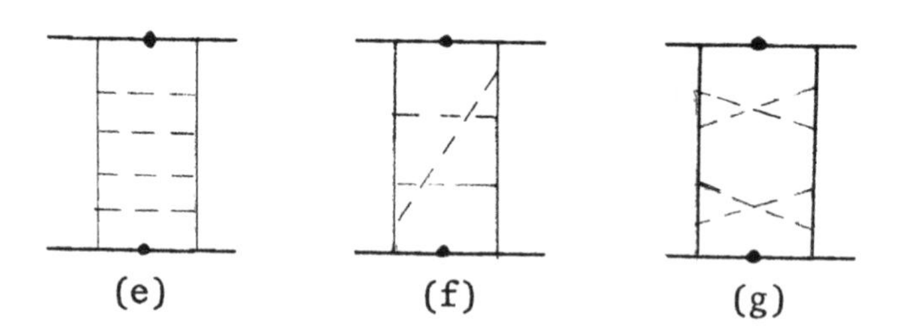

(e) (f) (g)

may be read off by inspection: $I_{(e)} \sim \ln^4 s$, $I_{(f)} \sim \ln^2 s$, $I_{(g)} \sim \ln^2 s$. For case (c), one has $I_{(c)} \sim \ln^3 s$; detailed calculation immediately gives $I_{(c)} = I_3 \simeq (2\pi i)^3 \cdot \frac{1}{3!} (\ln s/\mu^2)^3$.

The integrals defining (d) are obtained by adding another vertical NVM line, $G^{(2)}(\theta;q_3;-bp_2)$, in the manner of (10.22) and (10.23). The phase is now originally given by

$$\exp i[\xi_{12}k_1^{(-)}k_2^{(+)}+\eta_{12}K_1^{(-)}K_2^{(+)}+\theta_{12}Q_1^{(-)}Q_2^{(+)}],$$

with the appropriate associations:

$$K_1 \to -k_1,\ K_2 \to +k_3,\ Q_1 \to -k_3,\ Q_2 \to -k_2.$$

Upon performing $\int dk_1^{(-)}\int dk_2^{(-)}\int dk_3^{(-)}$, one finds

$$(2\pi i)^3\int_0^\infty \frac{dk_1^{(+)}}{k_1^{(+)}}\int_0^\infty \frac{dk_2^{(+)}}{k_2^{(+)}}\int_0^\infty \frac{dk_3^{(+)}}{k_3^{(+)}}\ \cdot$$

$$\cdot\ \exp\left[-i\left(\omega_1^2\,\frac{|\xi_{12}k_2^{(+)}-\eta_{12}k_3^{(+)}|}{k_1^{(+)}}+\omega_3^2\theta_{12}\,\frac{k_2^{(+)}}{k_3^{(+)}}\right)\right], \qquad (10.30)$$

where again the phase is always of the same sign and is small when $k_1^{(+)} > k_3^{(+)} > k_2^{(+)}$, leading to the same contribution as $I_{(c)}$. One difference between (10.30) and the corresponding calculation of $I_{(c)}$, is that the coefficient of $(k_1^{(+)})^{-1}$, in the phase of (10.30), can vanish for a fixed ratio $k_2^{(+)}/k_3^{(+)}$ and any $k_1^{(+)}$; however, at least in simple examples, this does not appear to effect the leading contribution. One should note that the graphs (h) and (d) are equivalent, but

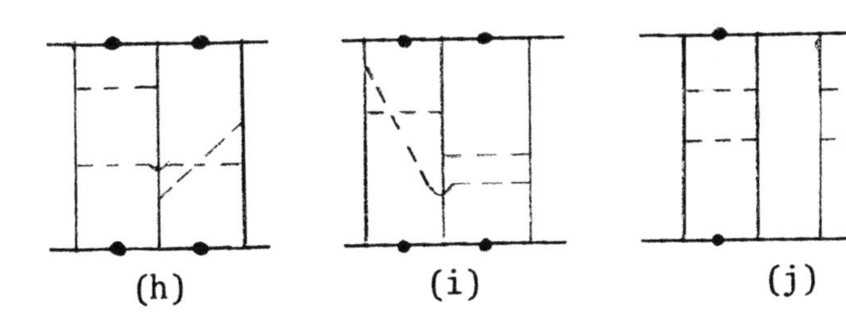

that (i) involves a real crossing of lines, and is down by one log.

From these examples, one may infer simple rules to extract the leading log dependence for such scalar pion exchange:

1) Consider only the ladder graphs, such as (a), (c), (d), for which the ordering of $k_\ell^{(+)}$ labels on every vertical line does not clash.

2) For any ladder graph of m horizontal lines so ordered, one has the leading log contribution $I_m \sim (2\pi i)^m \frac{1}{m!} (\ell n\, s/\mu^2)^m$, regardless of how the scalar pions are distributed between the vertical lines. For complicated nonplanar graphs, the validity of this last statement is not at all obvious; however, it will be assumed true in the approximate model of the next section.

There remains the nontrivial task of evaluating that $(\int d\xi_\ell \cdots \int d^2k_\ell \cdots)$ dependence multiplying the leading logs. This has been accomplished only for the ordinary ladder graphs, containing two vertical lines, or for their independent s-channel iterates, as in graph (j). From the above analysis it is straightforward to show that the leading log contribution to (10.17) consisting of the ordinary ladder graphs (n = 2) with m horizontal pions is given by

$$-\frac{g^4}{2}(2\pi)^{-4}\int \frac{d^2q_1}{q_1^2+m^2}\, e^{i\vec{q}_1\cdot\vec{b}} \int \frac{d^2q_2}{q_2^2+m^2}\, e^{i\vec{q}_2\cdot\vec{b}} \cdot \frac{1}{m!}\cdot$$

$$\cdot \left[\frac{\lambda^2}{8\pi}\,\bar{\alpha}([\vec{q}_1+\vec{q}_2]^2)\ \ell n\left(\frac{s}{\mu^2}\right)\right]^m ,$$

and hence the sum of all such terms, including m = 0, becomes

$$-2\left(\frac{g^2}{4\pi}\right)^2 \int d^2q\ e^{i\vec{q}\cdot\vec{b}}\ \bar{\alpha}(\vec{q}^2)\left(\frac{s}{\mu^2}\right)^{\frac{\lambda^2}{8\pi}\bar{\alpha}(\vec{q}^2)} , \qquad (10.31)$$

where

$$\bar{\alpha}(\vec{q}^2) = (2\pi)^{-2} \int d^2Q[Q^2+m^2]^{-1}[(Q-q)^2+m^2]^{-1}$$

$$= \frac{1}{4\pi} \int_0^1 dx[m^2+x(1-x)\vec{q}^2]^{-1}.$$

That term of (10.31) corresponding to $m = 0$ is simply the second s-channel iterate of $i\chi_1$ (given by (9.14)), $\frac{1}{2!}(i\chi_1)^2$, and appears in this expression because the expansion (10.17) includes disconnected graphs. It is clear from (10.14) and the previous analysis that $i\chi_2$ is given by (10.31) with the $m = 0$ contribution removed,

$$i\chi_2(s;b) = -2\left(\frac{g^2}{4\pi}\right)^2 \int d^2q \; e^{i\vec{q}\cdot\vec{b}} \; \bar{\alpha}(\vec{q}^2)[\left(\frac{s}{\mu^2}\right)^{\frac{\lambda^2}{8\pi}\bar{\alpha}(\vec{q}^2)} -1], \tag{10.32}$$

which, in the spirit of leading s-dependence, is equivalent to (10.31). These forms have the interpretation of simple Regge pole exchange (10.31), and multiple, s-channel Regge exchange (10.32), sometimes called the Regge-eikonal approximation.[1] For example, the amplitude obtained by the substitution of (10.31) into (9.13) is

$$T^{\text{Regge}}_{\text{Ladder}}(s;q^2) \sim i\left(\frac{g^2}{2m}\right)^2 \bar{\alpha}(\vec{q}^2)\left(\frac{s}{\mu^2}\right)^{1+\frac{\lambda^2}{8\pi}\bar{\alpha}(\vec{q}^2)}, \tag{10.33}$$

while the $i\chi_2$ of (10.32) leads to an eikonal amplitude of form

$$T^{\text{Regge}}_{\text{Eik.}}(s;q^2) \sim \frac{is}{2m^2}\int d^2b e^{i\vec{q}\cdot\vec{b}}[1-\exp\left(-\frac{as^{\alpha-1}}{\ell n\; s}\; e^{-cb^2/\ell n\; s}\right)], \tag{10.34}$$

where

$$\alpha = 1 + \frac{\lambda^2}{8\pi}\,\bar{\alpha}(0),$$

$$a = 2\left(\frac{g^2}{\lambda}\right)^2 \frac{\bar{\alpha}(0)}{|\bar{\alpha}'(0)|},$$

$$c = \frac{2\pi}{\lambda^2|\bar{\alpha}'(0)|},$$

and μ, the parameter which sets the scale for large s, has been taken as unity. In obtaining (10.34), the linear trajectory expansion $\bar{\alpha}(\vec{q}^{\,2}) \simeq \bar{\alpha}(0) + \vec{q}^{\,2}\,\bar{\alpha}'(0)$ has been used, valid for large impact parameters. While (10.33) will violate the Froissart bound, (10.34) will saturate it; the total cross section obtained from (10.34) is the same as that of the tower graph summations leading to (9.23), since the impact parameter integrand again cuts off for $b \sim \ell n\, s$. The absorptive parts of graphs (a) through (i) correspond to production in the multiperipheral manner. In particular, the total cross section for pion production may be obtained by summing (9.39) over all n, and making use of (10.11); in the leading log approximation this produces a total inelastic cross section quite similar to that obtained in the first paper of Reference 1.

C. A Limiting Model

Before one can take the result (10.34) seriously, an estimate of all the remaining $i\chi_n$, $n \geq 3$, must be made, and even in the context of summing leading logs this continues to be an unsolved problem. To indicate the types of structure which may appear, it is useful to invent a model which encompasses as much of the previous analysis as possible. It is somewhat simpler to give a complete functional formulation of this model, but the less elegant presentation used here will retain continuity with the discussion of the previous section.

Graph (d) is the simplest, connected, nonplanar contribution to χ_3, with a $\ell n\ s$ dependence estimated as in (10.30), but having a coefficient more complicated than that of the ordinary ladder graph (c). Where the latter had a coefficient proportional to $[\bar{\alpha}([\vec{q}_1+\vec{q}_2]^2)]^3$, one might expect the coefficient of (d) to be proportional to

$$\bar{\alpha}([\vec{q}_1+\vec{q}_2]^2)\bar{\alpha}([\vec{q}_2+\vec{q}_3]^2)\bar{\alpha}([\vec{q}_1+\vec{q}_3]^2),$$

but this is not the case. A model may be defined[2] by forcing this property, and in the following way. After all parametric ξ,η,θ integrals have been performed, integration over the transverse momenta (of the lowest rung of the graph) $\int d^2k_2$ yields a factor $\bar{\alpha}([\vec{q}_1+\vec{q}_3+\vec{k}_1-\vec{k}_3]^2)$. If the $\vec{q}_i^{\,2}$ are imagined to be larger than the $\vec{k}_j^{\,2}$, this may be replaced by $\bar{\alpha}([\vec{q}_1+\vec{q}_3]^2)$; and <u>then</u> the $\int d^2k_3$ may be performed to yield $\bar{\alpha}([\vec{q}_2+\vec{q}_3-\vec{k}_1]^2)$. Repeating the same approximation, $\int d^2k_1$ then produces $\bar{\alpha}([\vec{q}_1+\vec{q}_2]^2)$, and in this manner one arrives at the form analogous to that of the simpler ladder graphs.

One might imagine that summing on the other graphs of this form,

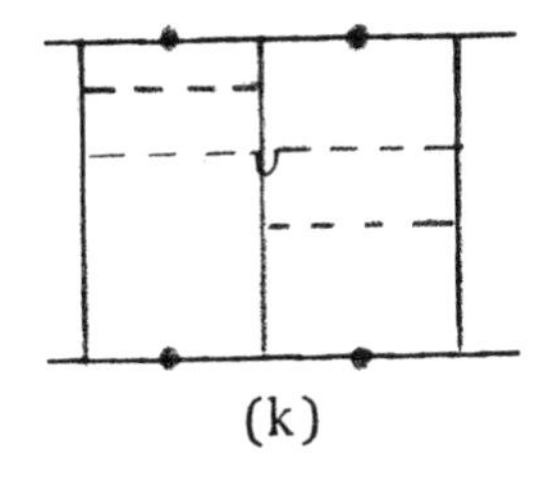
(k)

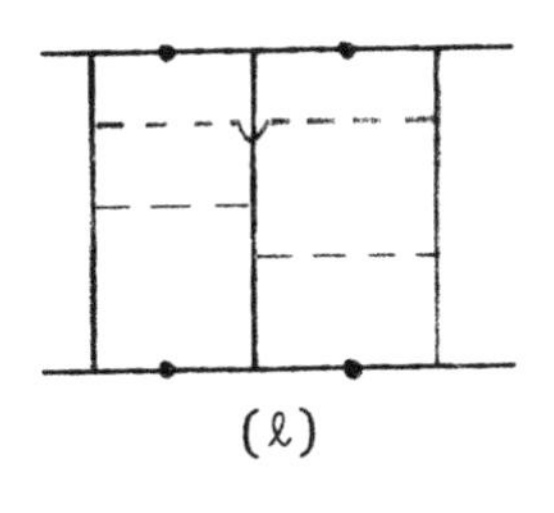
(ℓ)

would generate a simple result; but in fact the contributions of (k) and (ℓ) are identical to that of (d). In this connection, it might be useful to define "eikonal graphs," drawn in the form

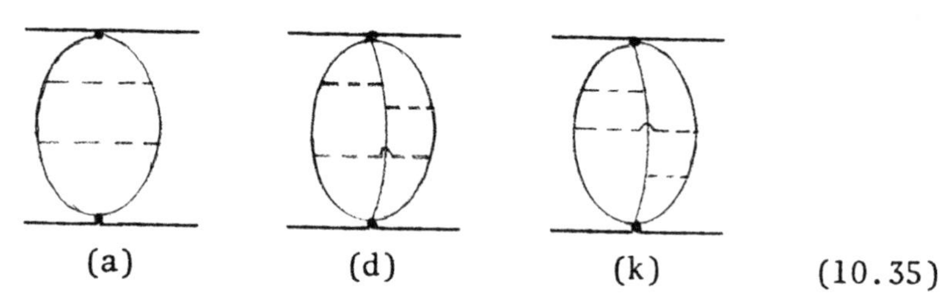

with labels indicating the contributions to which the graphs of (10.35) correspond. Such pictures make the "swivel invariance" of eikonal graphs obvious: one thinks of each NVM vertical line as a section of a hoop which may be rotated around the imaginary veritcal line joining the upper and lower dots of an eikonal graph. Swivel invariance, which may be demonstrated from the analytic forms defining each graph, states that the value of any eikonal graph is independent of how the hoops are swung about relative to each other. For example, (k) and (d) generate equal $\ell n\ s$ dependence,

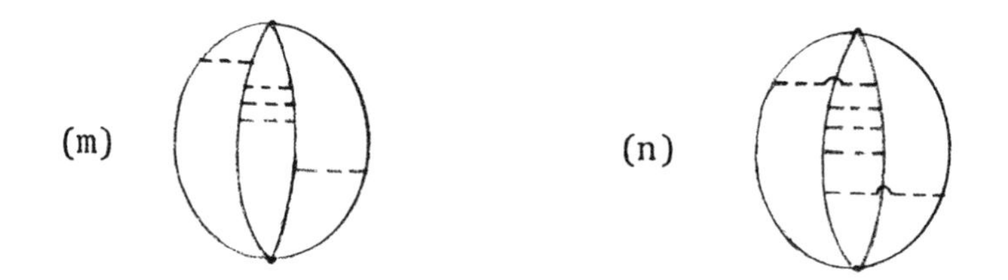

as do eikonal graphs (m) and (n).

Swivel invariance does not state that the leading log contributions of all eikonal graphs with the same number of vertical and horizontal lines is the same; but the limiting, eikonal graph model of this section will assume that property, assigning a factor $\bar{\alpha}_{ij} \equiv \bar{\alpha}([\vec{q}_i+\vec{q}_j]^2)$ whenever the NVM lines labeled by q_i and q_j are joined by a horizontal pion. This represents an extreme, or limiting situation, with minimum correlation between nonplanar pions. In accordance with the leading log estimate, each graph with m horizontal and n vertical lines is assumed to contribute to (10.17) the same $1/m!\ (\ell n\ s)^m$ dependence, multiplied by

$$\frac{(ig^2)^n}{n!}(2\pi)^{-2n}\int\frac{d^2q_1}{[q_1^2+m^2]}e^{i\vec{q}_1\cdot\vec{b}}\cdots\int\frac{d^2q_n}{[q_n^2+m^2]}e^{i\vec{q}_n\cdot\vec{b}}\;\cdot$$

$$\cdot\prod_{i\neq j}\bar{\alpha}_{ij}\cdot\left(\frac{\lambda^2}{8\pi}\right)^m. \tag{10.36}$$

There remains the question of counting the number of ways in which m horizontal lines may be distributed among n vertical lines (or hoops). For clarity and ease of notation, we first consider the case of three vertical lines, 1, 2, 3, joined by pion lines labeled n_{12}, n_{23}, and n_{13}. It is not difficult to see that the total number of ways of ordering $m = n_{12}+n_{23}+n_{13}$ horizontal lines is given by $(n_{12}+n_{13}+n_{23})!/n_{12}!n_{13}!n_{23}!$, which quantity must multiply (10.36), and yields

$$\frac{(ig^2)^3}{3!}(2\pi)^{-6}\int\frac{d^2q_1}{[q_1^2+m^2]}\int\frac{d^2q_2}{[q_2^2+m^2]}\int\frac{d^2q_3}{[q_3^2+m^2]}\;\cdot$$

$$\cdot e^{i(\vec{q}_1+\vec{q}_2+\vec{q}_3)\cdot\vec{b}}\cdot\frac{1}{n_{12}!}\left(\frac{\lambda^2}{8\pi}\bar{\alpha}_{12}\ \ell n\ \frac{s}{\mu^2}\right)^{n_{12}}\cdot$$

$$\cdot\frac{1}{n_{23}!}\left(\frac{\lambda^2}{8\pi}\bar{\alpha}_{23}\ \ell n\ \frac{s}{\mu^2}\right)^{n_{23}}\cdot\frac{1}{n_{13}!}\left(\frac{\lambda^2}{8\pi}\bar{\alpha}_{13}\ \ell n\ \frac{s}{\mu^2}\right)^{n_{13}}. \tag{10.37}$$

Summing over all n_{12}, n_{23}, n_{13}, one obtains

$$\frac{(ig^2)^3}{3!}(2\pi)^{-6}\prod_{n=1}^{3}\int\frac{d^2q_n}{[q_n^2+m^2]}e^{i\vec{q}_n\cdot\vec{b}}\;\cdot$$

$$\cdot \left(\frac{s}{\mu^2}\right)^{\frac{\lambda^2}{8\pi}[\bar{\alpha}_{12}+\bar{\alpha}_{23}+\bar{\alpha}_{13}]}, \tag{10.38}$$

which corresponds to Regge behavior between each pair of vertical NVM lines; in this sense, especially, the model represents a limiting situation. The corresponding contribution to $i\chi_3$ is obtained by subtracting from (10.38) $\frac{1}{3!}(i\chi_1)^3 + (i\chi_1)(i\chi_2)$.

The generalization of (10.38) to n vertical lines is simply

$$\frac{(ig^2)^n}{n!} \prod_{\ell=1}^{n} (2\pi)^{-2} \int \frac{d^2q_\ell}{[q_\ell^2+m^2]} e^{i\vec{q}_\ell\cdot\vec{b}} \cdot \left(\frac{s}{\mu^2}\right)^{\frac{\lambda^2}{8\pi}\sum_{i<j=1}^{n}\bar{\alpha}_{ij}}, \tag{10.39}$$

and there are two ingredients of this expression which are worth noting, (a) the $(i)^n$ factor which permits cancellations between successive, real terms, and (b) the s-dependence which grows rapidly with n, suggesting that such cancellations will be important. For computational simplicity we now make a sequence of approximations, beginning with the choice that all $\bar{\alpha}_{ij}$ be given by the same constant, $\lambda^2/8\pi\ \bar{\alpha}(0) = \alpha_0$; and then attempt to compute the sum over n of (10.39), or of the expression for σ_T given by (9.15), with $\exp(i\chi_0)-1$ replaced by (10.17),

$$\sigma_T = -\frac{4\pi}{m^2} \operatorname{Re} \sum_{n=2}^{\infty} c_n \cdot \frac{1}{n!} \left(\frac{ig^2}{4\pi}\right)^n \left(\frac{s}{\mu^2}\right)^{\alpha_0 n(n-1)/2}, \tag{10.40}$$

where $1/2\, n(n-1)$ represents the number of pairs of vertical lines, and

$$c_n = \int_0^\infty x dx [2K_0(x)]^n. \tag{10.41}$$

For large n, the important contributions to the integrand of (10.41) come from small x, and hence the approximation $K_o(x) \sim \theta(1-x)\ln(x-1)$ yields

$$\sigma_T \simeq -\frac{2\pi}{m^2} \operatorname{Re} \sum_{n=2}^{\infty} \left(\frac{ix}{y}\right)^n y^{n^2}, \tag{10.42}$$

where $x = g^2/4\pi$ and $y = (s/\mu^2)^{\alpha_o/2}$.

Under the assumptions of this analysis, $y > 1$, and (10.42) can hardly be considered a convergent series. One may attempt to give meaning to expressions such as these by considering x/y and $y < 1$, summing over all n, and then continuing to large y. Whatever the procedure, ingredients (a) and (b) make clear the possibility of cancellations and consequent deviations from (10.31) or (10.32), and hence from (9.23), which is the essential lesson to be learned from this model. As an example, the series of (10.42) may be represented in the form

$$S(x,y) = \frac{1}{\sqrt{\pi}} \int_{-\infty}^{+\infty} d\alpha\, e^{-\alpha^2} F\left(\frac{x}{y}\, e^{2\alpha\beta}\right), \quad \beta = \sqrt{\ln y}, \tag{10.43}$$

if

$$F(x) = \sum_{n=2}^{\infty} a_n (ix)^n, \tag{10.44}$$

in the sense that an expansion of (10.44) under the integral of (10.43) generates the sum

$$S(x,y) = \sum_n a_n \left(\frac{ix}{y}\right)^n y^{n^2}.$$

If (10.43) is assumed to be the proper definition of this sum for all y, the asymptotic estimate of σ_T may easily be performed.

The variable change $\ln u = -\beta^2 + 2\alpha\beta$ leads to the alternate representation

$$S(x,y) = \frac{1}{2\beta} e^{-\beta^2/4} \int_0^\infty \frac{du}{u^{3/2}} \exp[-\left(\frac{\ell n\, u}{2\beta}\right)^2] F(xu). \quad (10.45)$$

If the integral $\psi(x) = \int_0^\infty du \cdot u^{-3/2} F(xu)$ exists, then as $\beta \to \infty$,

$$S(x,y) \sim \frac{\psi(x)}{2\beta} e^{-\beta^2/4} \sim s^{-\alpha_0/8} (\ell n\, s)^{-1/2} \to 0,$$

providing an example of complete cancellation. This property is displayed by the present model, since $a_n = +1$, $-\mathrm{Re}F(x) = +x^2[1+x^2]^{-1}$, and the limit $\beta \to \infty$ may be taken under the integral of (10.45). Of course, this estimate of σ_T must not be taken seriously; it is merely indicative of the sort of behavior which can occur when the contributions of the remaining χ_n are included.

Such cancellations may exist for models involving less divergent sums than that of (10.40). For example, in a model in which Regge behavior exists only between nearest-neighbor NVM lines, (10.40) would be replaced by

$$\sigma_T = - \frac{4\pi}{m^2} \mathrm{Re} \sum_{n=2}^{\infty} \frac{c_n}{n!} \left(\frac{ig^2}{4\pi}\right)^n \left(\frac{s}{\mu^2}\right)^{\alpha_o(n-1)}, \quad (10.46)$$

and is a convergent series, for appropriate c_n, if the coupling g^2 is sufficiently small to satisfy $g^2/4\pi \cdot s/\mu^2 < 1$ while $s/\mu^2 > 1$. With

$$G(x) = \sum_{n=2}^{\infty} \frac{c_n}{n!} (ix)^n$$

one has

$$\sigma_T = -\frac{4\pi}{m^2}\left(\frac{s}{\mu^2}\right)^{-\alpha_o} \operatorname{Re} G\left(\frac{g^2}{4\pi}\cdot\left(\frac{s}{\mu^2}\right)^{\alpha_o}\right),$$

with a vast possibility of results as g^2 is increased. In statistical language, (10.40) corresponds to a strongly interacting system of particles, with forces independent of position, such that a thermodynamic limit does not exist, in contrast to the nearest-neighbor interactions to which (10.46) corresponds. In view of the current experimental efforts at high energies, a realistic estimate of such behavior is greatly to be desired.

Notes

1. S-J. Chang and T-M. Yan, Phys. Rev. D4, 537 (1971); G. M. Cicuta and R. L. Sugar, Phys. Rev. D3, 970 (1971); M. M. Islam, ICTP Preprint IC/71/80; A. R. Swift, "Theoretical Difficulties with the Regge-Eikonal Model," U. of Mass. Preprint (Nov. 1971).

2. The model-making of this section was done in collaboration with Professor R. Blankenbecler.

A. Ordered Exponentials

The time-ordered operator (2.1) has the expansion

$$T_b^a\{j\} = 1+i\int_{t_b}^{t_a} d^4x j(x)A(x) + \frac{i^2}{2!}\int_{t_b}^{t_a} d^4x_1\int_{t_b}^{t_a} d^4x_2 \; j(x_1)j(x_2)\cdot$$

$$\cdot\left(A(x_1)A(x_2)\theta(t_1-t_2)+A(x_2)A(x_1)\theta(t_2-t_1)\right)+ \cdots .$$

By symmetry, the term quadratic in the source $j(x)$ may be rewritten as

$$i^2\int_{t_b}^{t_a} d^4x_1 \int_{t_b}^{t_1} d^4x_2 \; j(x_1)j(x_2)A(x_1)A(x_2),$$

and in exactly the same way, that term with n factors of j may be rewritten so as to remove the $1/n!$ arising from the expansion of the exponential, but with the limits of $n-1$ integrals arranged in a time-ordered manner.

A simpler, but useful example of this equivalence follows from the solution to the elementary differential equation,

$$\frac{dF}{d\xi} = G(\xi)F(\xi),$$

under the simplest boundary condition, $F(0) = 1$. The solution is an ordinary exponential only if $[G(\xi_1),G(\xi_2)] = 0$. If $G(\xi)$ is an arbitrary operator, such that this commutator does not vanish, the solution is an ordered exponential,

$$F(\xi) = \left(\exp\int_0^\xi d\xi' G(\xi')\right)_+,$$

with the ordering symbol referring to the dummy ξ'

variables, so that the $G(\xi')$ of largest ξ' stand to the left, in any expansion of the exponential.

That this is indeed the proper solution may be verified by differentiation. One differentiates an ordered quantity in the usual way, "bringing down" the result of differentiation and placing the latter anywhere inside the ordered bracket, since it is the ordering symbol which states precisely how the terms are to be arranged upon subsequent expansion. One may therefore write

$$\frac{d}{d\xi}\left(\exp\int_0^{\xi} d\xi' G(\xi')\right)_+ = \left(G(\xi)\left[\exp\int_0^{\xi} d\xi' G(\xi')\right]\right)_+$$

$$= \left(\left[\exp\int_0^{\xi} d\xi' G(\xi')\right]G(\xi)\right)_+ .$$

Upon expansion of the exponent, however, every $\xi' \leq \xi$, and hence the $G(\xi)$ factor must always be placed on the extreme LHS, and is equivalent to

$$G(\xi)\left(\exp\int_0^{\xi} d\xi' G(\xi')\right)_+ .$$

The alternate expression is most simply obtained by combining differential equation and boundary condition into an integral equation,

$$F(\xi) = 1 + \int_0^{\xi} d\xi' G(\xi') F(\xi').$$

Iteration produces the sequence of terms with ordered limits of integration,

$$F(\xi) = 1 + \int_0^{\xi} d\xi' G(\xi') + \int_0^{\xi} d\xi_1 \int_0^{\xi_1} d\xi_2 G(\xi_1) G(\xi_2) + \cdots ,$$

demonstrating the equivalence of this form to the expansion of the ordered exponential.

Just as the solution to this ordinary differential equation is not simply an ordinary exponential, so the derivative of an exponential is not simply given by the product of derivative times function,

$$\frac{d}{d\xi} \exp H(\xi) \neq \frac{dH}{d\xi} \exp H(\xi).$$

The proper definition may be obtained by introducing the quantity

$$Q(\lambda) = \exp[-\lambda H]\cdot\exp[\lambda(H+\delta H)], \quad Q(0) = 1,$$

and calculating its derivative with respect to λ,

$$\frac{dQ}{d\lambda} = \exp[-\lambda H]\cdot\delta H\cdot\exp[+\lambda H]\cdot Q(\lambda).$$

Hence, to first order in δH,

$$Q(\lambda) \sim 1 + \int_0^{\lambda} d\lambda' \exp[-\lambda' H]\delta H \exp[\lambda' H],$$

and a comparison with the definition of $Q(1)$ yields

$$\frac{d}{d\xi} \exp H(\xi) = \int_0^1 d\lambda \exp[(1-\lambda)H(\xi)]\cdot\frac{dH}{d\xi}\cdot\exp \lambda H(\xi)$$

$$= \int_0^1 d\lambda \exp[\lambda H(\xi)]\cdot\frac{dH}{d\xi}\cdot\exp[(1-\lambda)H(\xi)],$$

which reduces to the familiar form only if $[H, dH/d\xi] = 0$. If $H = \int_0^{\xi} d\xi' G(\xi')$, then

$$\frac{d}{d\xi} \exp \int_0^{\xi} d\xi' G(\xi') = \int_0^1 d\lambda \exp[(1-\lambda)\int_0^{\xi} d\xi' G]\cdot G(\xi)\cdot$$

$$\cdot \exp[\lambda \int_0^\xi d\xi' G],$$

and therefore

$$\frac{d}{d\xi} \left(\exp \int_0^\xi d\xi' G(\xi')\right)_+ = G(\xi) \left(\exp \int_0^\xi d\xi' G(\xi')\right)_+,$$

as stated above.

B. Functional Derivative

A definition of functional differentiation is given by

$$\frac{\delta}{\delta j(x)} F\{j(u)\} = \lim_{\varepsilon \to 0} \frac{1}{\varepsilon} [F\{j(u)+\varepsilon\delta^4(x-u)\}-F\{j(u)\}],$$

where $F\{j(u)\}$ denotes an arbitrary functional of j, with coordinate u--which is integrated over in some arbitrarily complicated way in the definition of $F\{j\}$ --written explicitly. An elementary example is

$$\frac{\delta}{\delta j(x)} \exp \int d^4z\; f(z)j(z) = f(x)\cdot \exp \int d^4z\; f(z)j(z).$$

C. Inconsistency of the Strong-Asymptotic Condition

Consider the commutator built out of the A_{IN} fields,

$$\Delta(x-y) = -i\langle 0|[A_{IN}(x),A_{IN}(y)]|0\rangle,$$

and the commutator built out of the dressed, fully interacting fields $A(x)$,

$$\Delta'(x-y) = -i\langle 0|[A(x),A(y)]|0\rangle.$$

Were the strong, or operator asymptotic condition valid,

$$A(x) \to Z_3^{1/2} A_{IN}(x), \quad x_0 \to -\infty,$$

one could let both x and y recede to $-\infty$ keeping their difference fixed, and obtain $\Delta'(x-y) = Z_3\Delta(x-y)$, a relation which has as its consequence (compare (5.21)) the absence of all interaction.

D. The Baker-Hausdorf Formula

This most useful relation states that if A and B are operators, but their commutator $[A,B]$ is a c-number, then

$$\exp[A+B] = \exp A \cdot \exp B \cdot \exp(-\frac{1}{2}[A,B]),$$

or equivalently,

$$\exp A \cdot \exp B = \exp B \cdot \exp A \cdot \exp[A,B].$$

The derivation proceeds as in Appendix A, by defining an operator

$$F(\lambda) = \exp[-\lambda A] \cdot \exp[\lambda(A+B)],$$

and calculating its differential equation,

$$F'(\lambda) = \exp[-\lambda A] \cdot B \cdot \exp[\lambda A] \cdot F(\lambda).$$

Because the commutator $[A,B]$ is assumed to commute with A, the expansion of the RHS factor multiplying $F(\lambda)$ has but two terms, $e^{-\lambda A}Be^{\lambda A} = B-\lambda[A,B]$. Further, because $[B,[A,B]] = 0$, $F'(\lambda)$ may be integrated directly to give

$$F(\lambda) = \exp[\lambda B - \frac{\lambda^2}{2}[A,B]],$$

which, upon setting $\lambda = 1$, produces the desired formulae. Expanding to terms linear in A yields

$$[A,e^B] = [A,B]e^B,$$

which has been used in (2.22) and (3.14).

E. Absence of a Lower Bound

Let $\pi(x) = \partial_0 A(x)$ be the momentum operator conjugate to the field $A(x)$, and define the time-dependent unitary operator

$$U(t) = \exp[i\int d^3x\ \chi(\vec{x},t)\pi(\vec{x},t)],$$

where $\chi(x)$ denotes an arbitrary but definite c-number function. The time-independent Hamiltonian of this problem is $H = H_0 + H_1$, where

$$H_0 = \frac{1}{2}\int d^3x\left((\vec{\nabla}A)^2 + \mu^2A^2 + \pi^2\right)$$

and $H_1 = g/n! \int d^3xA^n$. Under this unitary transformation, $A \to A' = UAU^\dagger = A(x) + \chi(x)$, a result following from the ETCR. The new interaction Hamiltonian is

$$H_1' = \frac{g}{n!}\int d^3x[A+\chi]^n,$$

with a similar form for H_0. For odd $n \geq 3$, the choice $\chi \to -\infty$ generates an increasingly negative contribution to H_1', and therefore to H', since H_0' is not so effected. Thus the eigenstates of this theory may be expected to have no lower bound. The original argument is due to G. Baym, Phys. Rev. 117, 886 (1960).

F. The Gaussian Combinatoric

This derivation follows that given by C. Sommerfield, Ann. of Phys. 26, 1 (1963). It appears in this form in the unpublished lecture notes of B. Zumino (New York University, 1958). An alternate version was given by J. Schwinger, Stanford Summer Lectures (1956).

Define

$$F(\lambda) = \exp[-\frac{i}{2}\lambda\int\frac{\delta}{\delta j}A\frac{\delta}{\delta j}]\cdot\exp[\frac{i}{2}\int jBj],$$

and adopt the ansatz

$$F(\lambda) = \exp M(\lambda),$$

$$M(\lambda) = \frac{i}{2}\int j\chi(\lambda)j + L(\lambda),$$

with $\chi(x,y|\lambda)$ and $L[\lambda]$ independent of $j(x)$. Substituting into the differential equation

$$F'(\lambda) = -\frac{i}{2}\int \frac{\delta}{\delta j} A \frac{\delta}{\delta j} \cdot F(\lambda),$$

and equating coefficients of j, leads to the pair of relations

$$\frac{d\chi(u,v|\lambda)}{d\lambda} = \int dx \int dy \chi(u,x|\lambda)A(x,y)\chi(y,v|\lambda),$$

or $d\chi/d\lambda = \chi A\chi$, and $dL/d\lambda = \frac{1}{2}\,\mathrm{Tr}[A\chi]$. With the obvious boundary conditions, $L(0) = 0$, $\chi(u,v|0) = B(u,v)$, there follow the solutions

$$\chi(\lambda) = B[1-\lambda AB]^{-1} = [1-\lambda BA]^{-1}B,$$

and

$$L(\lambda) = \frac{1}{2}\,\mathrm{Tr}\,\ell n[1-\lambda AB]^{-1},$$

which yield (3.56).

G. A Matrix Relation

The relation $\exp[\mathrm{Tr}\,\ell nM] = \det M$ may be derived for a finite, hermitian matrix M by writing $M = 1+\chi$ and assuming that $\chi' = S^{\dagger}\chi S$ is diagonal. Then

$$\det(1+\chi) = \det(1+S\chi'S^{\dagger}) = \det(1+\chi') = \prod_i (1+\chi_i'),$$

where $\chi'_{ij} = \delta_{ij}\chi'_i$. By inspection, this is the same as

$$\exp[\mathrm{Tr}\ \ell n(1+\chi)] = \exp[\mathrm{Tr}\ \ell n(1+\chi')],$$

since

$$\mathrm{Tr}\ \ell n(1+\chi') = \mathrm{Tr}\sum_{n=1}^{\infty} \frac{(-1)^{n-1}}{n} (\chi')^n = \sum_{n=1}^{\infty} \frac{(-1)^{n-1}}{n} \sum_i (\chi'_i)^n =$$

$$= \sum_i \ell n(1+\chi'_i).$$

Application to Chapter 4 involves the quantity

$$\frac{1}{2}\mathrm{Tr}\ \ell n F = \frac{1}{2}\int_0^1 d\lambda\ \mathrm{Tr}(F[1+\lambda F]^{-1}).$$

Following the derivation of Appendix F, the Trace operation is to be taken over all coordinates, continuous and discrete, so that this quantity becomes

$$\frac{1}{2}\int_0^1 d\lambda \int dx \sum_a \langle x|\overline{F}[1+\lambda\overline{F}]^{-1}|x\rangle_{aa},$$

where $\underline{a}$ denotes the isotopic indices. With $\langle x|F|y\rangle_{ab} = \delta^4(x-y)\overline{F}(\pi(x))_{ab}$, one obtains

$$\frac{1}{2}\delta^4(0)\int d^4x \int_0^1 d\lambda \sum_a \left(\overline{F}(x)[1+\lambda\overline{F}(x)]^{-1}\right)_{aa}$$

$$= \frac{1}{2}\delta^4(0)\int d^4x\ \mathrm{tr}[\ell n F(x)],$$

where the trace operation now represents a sum over isotopic coordinates only. Using the general relation of this appendix, one obtains the stated result,

$\frac{1}{2}\,\delta^4(0)\int d^4x\ \ell n[\det F(x)].$

H. A Reciprocity Relation

Let $F\{A\}$ be given by any sum of polynomials, most simply represented by $\exp[\int fA]$, with $f(z)$ arbitrary and susceptible to functional differentiation. Then it immediately follows that

$$F\left\{\frac{1}{i}\frac{\delta}{\delta j}\right\}\exp[\tfrac{i}{2}\int j\Delta_c j] = \exp[\tfrac{i}{2}\int j\Delta_c j]\cdot$$

$$\cdot\exp[-\tfrac{i}{2}\int \frac{\delta}{\delta A}\Delta_c\frac{\delta}{\delta A}]F\{A\},$$

if $A \equiv \int \Delta_c j$.

I. A Linkage Relation

Represent $G_I[A]$ and $G_{II}[A]$ by arbitrary sums of polynomials, in terms of two functions $f_I(z)$, $f_{II}(z)$, each of which is susceptible to functional differentiation. Then,

$$\exp[-\tfrac{i}{2}\int \frac{\delta}{\delta A}\Delta_c\frac{\delta}{\delta A}]\exp\int[f_I+f_{II}]A$$

$$= \exp\int[f_I+f_{II}]A\cdot\exp[-\tfrac{i}{2}\int(f_I+f_{II})\Delta_c(f_I+f_{II})].$$

The cross terms of the second RHS factor correspond to linkages between G_I and G_{II}, and one immediately obtains (3.67). The general statement, for many factors $G_i[A]$, may be derived in the same way,

$$\exp[-\tfrac{i}{2}\int \frac{\delta}{\delta A}\Delta_c\frac{\delta}{\delta A}]\prod_{i=1}^{n} G_i[A] =$$

$$= \exp[-i \sum_{i>j=1}^{n} \int \frac{\delta}{\delta A_i} \Delta_c \frac{\delta}{\delta A_j}] \cdot$$

$$\cdot \prod_{i=1}^{n} (\exp[- \frac{i}{2} \int \frac{\delta}{\delta A_i} \Delta_c \frac{\delta}{\delta A_i}] \cdot G_i[A_i]) \Big|_{A_i = A} .$$

J. A Gauge Formula

Again, let $G_F[A]$ be represented by $\exp\int f_\mu A_\mu$, with $f_\mu(z)$ susceptible to functional differentiation, thereby producing the expansions of $G_F[A]$. Then, with one integration-by-parts,

$$\frac{\delta}{\delta \Lambda(z)} G_F[A+\partial\Lambda] = -\partial_\mu^z \frac{\delta}{\delta A_\mu(z)} G_F[A+\partial\Lambda],$$

for arbitrary $\Lambda(z)$. But, by construction, $G_F[A+\partial\Lambda]$ is independent of Λ; and hence

$$\partial_\mu^z \frac{\delta}{\delta A_\mu(z)} G_F[A] = 0,$$

as stated.

K. Gaussian Functional Integral

The art of evaluating integrals without integration is displayed perhaps most elegantly for Gaussians, such as

$$Q_\pm[j;B] = \int d[\chi] \exp[i \int j\chi \mp \frac{i}{2} \int \chi \cdot B \cdot \chi].$$

By inspection,

$$\exp[\pm \frac{i}{2} \int \frac{\delta}{\delta j} A \frac{\delta}{\delta j}] Q_\pm[j;B] = Q_\pm[j;B+A]. \tag{K1}$$

Choose the ansatz $Q_{\pm}[j;B] = N_{\pm}[B]\exp[\pm \frac{i}{2} \int j \cdot B^{-1} \cdot j]$, and substitute into (K1). From the result of Appendix F, one sees that the j-dependence on both sides of the resulting equation is the same, and that

$$N_{\pm}[B]e^{\frac{1}{2} \mathrm{Tr}\ \ell n B} = N_{\pm}[A+B]e^{\frac{1}{2} \mathrm{Tr}\ \ell n [A+B]} . \qquad \text{(K2)}$$

Since (K2) is independent of A, it is also independent of B, and is therefore a constant, $C_{\pm}$, from which follows (4.14).

L. The Connectedness Lemma

The statement (6.2) may be demonstrated by the following construction based upon the simplest self-interaction of a boson field, although both method and result are more generally valid. For the interaction Lagrangian $L' = -\ g/n!\ A^n$ one has

$$\exp[-\frac{i}{2} \int j \cdot \Delta_c \cdot j] \cdot NZ\{j\}$$

$$= \exp[-\frac{i}{2} \int \frac{\delta}{\delta A} \Delta_c \frac{\delta}{\delta A}] \cdot \exp L[A] \equiv S[A],$$

where $A \equiv \int \Delta_c j$, and $L[A] = -\ i/n!\ g \int A^n$. For any $L[A]$, the RHS $S[A]$ is given by the exponential of connected quantities, where "connected" means linked by at least one virtual Δ_c propagator.

One defines the connected quantities

$$Q_N[A] = [\exp\left(-\frac{i}{2} \int \frac{\delta}{\delta A} \Delta_c \frac{\delta}{\delta A}\right) \cdot L^N[A]]_{\mathrm{conn}}$$

$$= \exp[-i \sum_{i>j}^{N} \int \frac{\delta}{\delta A_i} \Delta_c \frac{\delta}{\delta A_j}] Q_1[A_1] \cdots Q_1[A_N] \Bigg|_{\mathrm{conn}}^{A_i = A},$$

where

$$Q_1[A] = \exp[-\frac{i}{2}\int \frac{\delta}{\delta A}\Delta_c \frac{\delta}{\delta A}]L[A].$$

Then,

$$S = \sum_{N=0}^{\infty} \frac{1}{N!} \exp[-\frac{i}{2}\int \frac{\delta}{\delta A}\Delta_c \frac{\delta}{\delta A}]L^N[A]$$

$$= \sum_{N=0}^{\infty} \frac{1}{N!} \exp[-i\sum_{i>j}^{N}\int \frac{\delta}{\delta A_i}\Delta_c \frac{\delta}{\delta A_j}]Q_1[A_1]\cdots Q_1[A_N]\Big|_{A_i=A}.$$

In any array of N such Q_1 factors, one factor-pairs in all possible ways, grouping m_1 Q_1s singly, $2m_2$ Q_1s in pairs to form m_2 terms Q_2, $3m_3$ Q_1s to form m_3 Q_3s, etc. The number of ways of dividing up N such factors is $N![m_1!(2m_2)!(3m_3)!\cdots(nm_n)!]^{-1}$ while the number of Q_2 pairs made is $(2m_2)!/m_2!(2!)^{m_2}$, the number of Q_3s formed is $(3m_3)!/m_3!(3!)^{m_3}$, etc. Hence the contribution to the sum made by the N*th* term is

$$\frac{1}{N!} \times N![m_1!(2m_2)!\cdots(nm_n)!]^{-1} \times$$

$$\times \frac{(2m_2)!}{m_2!(2!)^{m_2}} \times \cdots \times \frac{(nm_n)!}{m_n!(n!)^{m_n}} \times Q_1^{m_1} Q_2^{m_2}\cdots Q_n^{m_n}.$$

Since the sum of N runs from 0 to ∞, with $N = m_1 + 2m_2 + 3m_3 + \cdots + nm_n$, one may sum on each m_i separately to obtain

$$S = \exp[Q_1 + \frac{1}{2!} Q_2 + \frac{1}{3!} Q_3 + \cdots],$$

which proves the assertion.